GUIDE TO THE USE OF THE WIND LOAD PROVISIONS

of ASCE 7-98

Kishor C. Mehta
Dale C. Perry

Abstract: The objective of the *Guide to the Use of Wind Load Provisions of ASCE 7-98* (formerly ASCE 7-88) is to provide guidance in the use of the wind load provisions set forth in ASCE Standard 7-98, *Minimum Design Loads for Buildings and Other Structures*. In order to clearly identify the scope and limitations of this Standard, the *Guide* first provides a brief review of the background material that forms the basis for the Standard's provisions. It then discusses the general format of an analytical procedure used to determine wind loads and the various wind load parameters involved in this determination, such as velocity pressure, gust response factor, and pressure coefficients. This discussion is followed by examples using this analytical procedure to determine wind load. Finally, the *Guide* presents additional background information on the different wind parameters and a discussion on using the wind-tunnel procedure to determine wind load.

Library of Congress Cataloging-in-Publication Data

Mehta, Kishor C.
    Guide to the use of wind load provisions of ASCE 7-98 / Kishor C. Mehta, Dale C. Perry.
       p. cm.
    Rev. ed. of: Guide to the use of wind load provisions of ASCE 7-95. c1998.
    Includes bibliographical references and index.
    ISBN 0-7844-0533-6
    1. Wind-pressure. 2. Structural engineering. 3. Standards, engineering—United States.
4. Buildings—Standards—United States. I. Perry, Dale C. II. Mehta, Kishor C. Guide to the use of wind load provisions of ASCE 7-95. III. Title.

TH891 .M45 2001
624. 1'72'021873—dc21   2001055233

In Memoriam

Guide to the Use of the Wind Load Provisions of ASCE 7-98 was assembled as a combined effort of the two authors. Prior to its final editing, Dale C. Perry passed away in an accident while he was inspecting a damaged building, a task that he loved to do. The completed *Guide* is dedicated to the memory of my colleague and friend of twenty-five years.

Kishor C. Mehta

Preface

This guide is designed to assist professionals in the use of the wind load provisions of the American Society of Civil Engineers (ASCE) Standard ASCE 7-98. The *Guide* is a revision of *Guide to the Use of Wind Load Provisions of ASCE 7-95* because the wind load provisions underwent significant changes from the previous Standard, ASCE 7-95. The *Guide* contains 10 example problems worked out in detail, which can provide direction to practicing professionals in assessing wind loads on a variety of buildings and other structures. Every effort has been made to make these illustrative example problems correct and accurate. The authors would be pleased to receive comments regarding inaccuracies, errors, or different interpretations. The views expressed and interpretation of the wind load provisions made in the *Guide* are those of the authors and not of the ASCE 7 Standards Committee or the ASCE organization.

Authors' Disclaimer

Although the authors have done their best to ensure that any advice, recommendation, interpretation, or information given herein is accurate, no liability or responsibility of any kind (including liability for negligence) is accepted by the authors.

Acknowledgments

The authors wish to acknowledge the members of the ASCE 7 Standards Committee that was chaired by Dr. Jim Harris during the consensus process of ASCE 7-98. The members of the Wind Load Task Committee and the ASCE 7 Standards Committee contributed significantly to the final wind load provisions of ASCE 7-98 through their questions, comments, and discussions. The authors are indebted to these 100 plus members.

In a document of this type, there are individuals in the background who helped in layout, word-processing, and checking calculations. These tasks were handled by the staff of the Wind Science and Engineering Research Center, Texas Tech University. Personnel at Texas A&M University also assisted in formulating the example problems. Contributions by these individuals are acknowledged by the authors.

About the Authors

Kishor C. Mehta, P.E., Honorary Member of ASCE, Horn Professor of Civil Engineering, is the Director of the Wind Science and Engineering Research Center at Texas Tech University, Lubbock, Texas. He served as Chairman of the ASCE 7 Task Committee on Wind Loads, which produced ASCE 7-88 and ASCE 7-95. He was lead author of the *Guide to the Use of Wind Load Provisions of ASCE 7-95*. Dr. Mehta is past president of the American Association of Wind Engineering and past chairman of the Committee on Natural Disasters, National Research Council. He is project director of the NSF-sponsored Colorado State University/Texas Tech University Cooperative Program in Wind Engineering and program director of the Texas Tech/National Institute of Standards and Technology Cooperative Agreement for Windstorm Damage Mitigation. In April 2000, the National Hurricane Conference honored Dr. Mehta with an award for distinguished service in wind engineering.

Dale C. Perry (1932–2001) was a Dockery Professor in the Department of Architecture at Texas A&M University, College Station, Texas. Dr. Perry worked for the Southern Building Code Congress International (secretary to the Wind Loads Deemed-to-Comply Ad Hoc Committee); served on ASCE/ANSI, SBCCI, and ICBO in developing performance criteria for wind/environmental loads; and was director of Research and Engineering for the Metal Building Manufacturers Association. He was past president of the American Association of Wind Engineering. He served on the ASCE Task Committee to Mitigate Wind Damage and on the ASCE 7 Task Committee on Wind Loads. He also served as a team leader or as a member of a number of post-disaster investigations, including Hurricanes Elena, Andrew, Iniki, Iwa, Gilbert, and Georges. He was the recipient of the engineering award of the National Hurricane Conference in 1993 and again in 1997.

Contents

List of Tables

List of Figures

Chapter 1.
INTRODUCTION

The American Society of Civil Engineers (ASCE) publication *Minimum Design Loads for Buildings and Structures, ASCE 7-98*, is a consensus standard. Its origin goes back to 1972 when the American National Standards Institute published a standard with the same title (ANSI A58.1-1972). That 1972 standard was revised 10 years later containing an innovative approach to wind loads for components and cladding of buildings (ANSI A58.1-1982). Wind load specifications were based on the understanding of aerodynamics of wind pressures in building corners, eaves, and ridge areas, as well as the effects on pressures of area averaging.

In the mid-1980s, the ASCE assumed responsibility for the committee, which establishes design loads for buildings and other structures. The document published by ASCE (ASCE 7-88) contained design load criteria for live loads, snow loads, wind loads, earthquake loads, and other environmental loads, as well as load combinations. The ASCE 7 Committee on Minimum Design Loads has voting membership of close to 100 individuals representing all aspects of the building construction industry. The criteria for each of the environmental loads are developed by respective task committees.

The wind load criteria of ASCE 7-88 (ASCE, 1990) were essentially the same as ANSI A58.1-1982. In 1995, ASCE published ASCE 7-95, which contained major changes in wind load criteria because the basic wind speed averaging time was adopted as 3-second gust. This change in wind speed averaging time from fastest-mile to 3-second gust necessitated significant changes in boundary-layer profile parameters, gust effect factor, and some pressure coefficients. A *Guide to the Use of the Wind Load Provisions of ASCE 7-95* (Mehta and Marshall, 1998) was published by ASCE to assist practicing professionals in the use of wind load criteria of ASCE 7-95.

In 2000, ASCE published a revision of ASCE 7-95 with updated wind load provisions. The document is termed ASCE 7-98 with the same title (ASCE, 2000). This Guide is designed to assist practicing professionals in the use of wind load criteria of ASCE 7-98.

1.1 OBJECTIVE OF THE GUIDE

The objective of this Guide is to provide direction in the use of wind load provisions of ASCE 7-98 (referred to as the Standard). The commentary of ASCE 7-98 Section C6 contains a good background and discussion of the wind load criteria; that information is not repeated here. Rather, this Guide contains two important items that will assist the users of ASCE 7-98: (1) Examples, and (2) Frequently Asked Questions.

This Guide contains 10 worked examples. Sufficient details of calculation of wind loads are provided to help the reader properly interpret the wind load provisions of the Standard. Section 6 of the Standard, as well as the figures and tables of the Standard, are cited liberally in the examples. To avoid repeating the words "ASCE 7-98" or "the Standard," the section, table, and figure numbers are given without additional references. **It is necessary to have a copy of ASCE 7-98 to follow the examples and work with this Guide.** A copy of ASCE 7-98 can be ordered by calling 1-800-548-ASCE or order on the web at *www.pubs.asce.org*.

1.2   SIGNIFICANT CHANGES

The wind load provisions of Section 6 were completely reorganized in ASCE 7-98 in an attempt to make them more consistent with the formats used by the model code organizations. The major change involved the replacement of Table 6-1 in ASCE 7-95 (and previous versions of the Standard) that provided the appropriate equations for the analytical procedure. This change necessitated separate provisions be given for each building and other structure categories. A simplified method is formulated to facilitate the assessment of wind loads for simple diaphragm buildings.

The basic approach to assessing wind loading has not been changed, but new sections have been added to clarify and simplify the applications of terrain and height factors, gust effect factors, enclosure classifications, and internal and external pressure coefficients. Updated wind speed contours have been added to reflect a more complete analysis of hurricane wind speeds.

The significant changes are listed below:

- **Method 1 – Simplified Procedure** provides tables of design pressures for MWFRS (main wind-force resisting systems) and C&C (components and cladding) for simple diaphragm buildings of regular shape, with heights of 30 ft or less, having enclosed and partially enclosed classifications. The design pressures are given for Exposure B with multiplying factors to convert pressures to other exposure categories.

- **Method 2 – Analytical Procedure** remains essentially unchanged from ASCE 7-95 with the exception of an additional directionality factor and that appropriate site exposure categories may be used for both **rigid buildings of all heights** and **low-rise buildings** for both MWFRS and C&C (exposure categories K_z have been revised accordingly).

- **Method 3 – Wind Tunnel Procedure** remains essentially the same with some new restrictions on use.

- New definitions ("approved," "building envelope," "impact resistant glazing," "escarpment," "hill," "hurricane-prone region," "impact resistant covering," "mean roof height," "openings," "regular shaped building (conditions for Method 1)," "ridge," "simple diaphragm building (conditions for Method 1)," "wind-borne debris region" have been added to clarify application of the provisions.

- Hurricane contours have been updated based upon more complete data analysis and selection of a "limit state return period" of approximately 500 years reduced to a "modified" 50-year return period.

- A comparison of the hurricane region contours of ASCE 7-95 and ASCE 7-98 is shown in Figures 1.1 through 1.3 and more appropriately reflect the attenuation of hurricane wind speeds inland after landfall.

- A directionality factor K_d is added into the equation for velocity pressure. The addition of the directionality factor necessitated changing the load factor of 1.3 to 1.6 in Section 2.

- New factors to convert 50-year speeds to mean recurrence intervals of 5-year to 500-year are given in the Commentary.

- Terrain exposure categories have been redefined to clarify their use for a specific design application. Aerial photographs have been added to the Commentary for clarification of exposures.

- New exposure tables K_z are provided to reflect the use of appropriate site terrain exposure categories for all MWFRS and C&C load calculations for analytical procedures (including low-rise buildings). The 0.85 factor in ASCE 7-95 for Exposure B has been removed.

- Terrain Exposure Category D is now excluded for the hurricane-prone coastal regions reflecting the observed increased roughness of ocean surfaces produced by hurricane events.

- The topographic effects section has been rewritten to add equations to the previous figure and reduce the height of the topographic feature for which the coefficient K_{zt} applies.

- Equations for the evaluation of the gust effect factor G have been added including a default value of 0.85 for rigid structures, irrespective of terrain exposure.

- A new section on enclosure classifications has been added to assist in the application of internal pressure coefficients.

- The internal pressure coefficients GC_{pi} have been revised for partially enclosed buildings with a reduction factor for large volume buildings.

- The figure for external pressure coefficients C_p for buildings of all heights has been expanded to include mansard and monoslope roofs.

- Added limitations are given on the use of wind tunnel testing.

Changes in tables and figures and in the wind tunnel procedure incorporate the latest available technical information. As noted above, the basic methodology of the Standard remains the same as in ASCE 7-95. Additional information on the changes can be found in the Commentary of the Standard and from references.

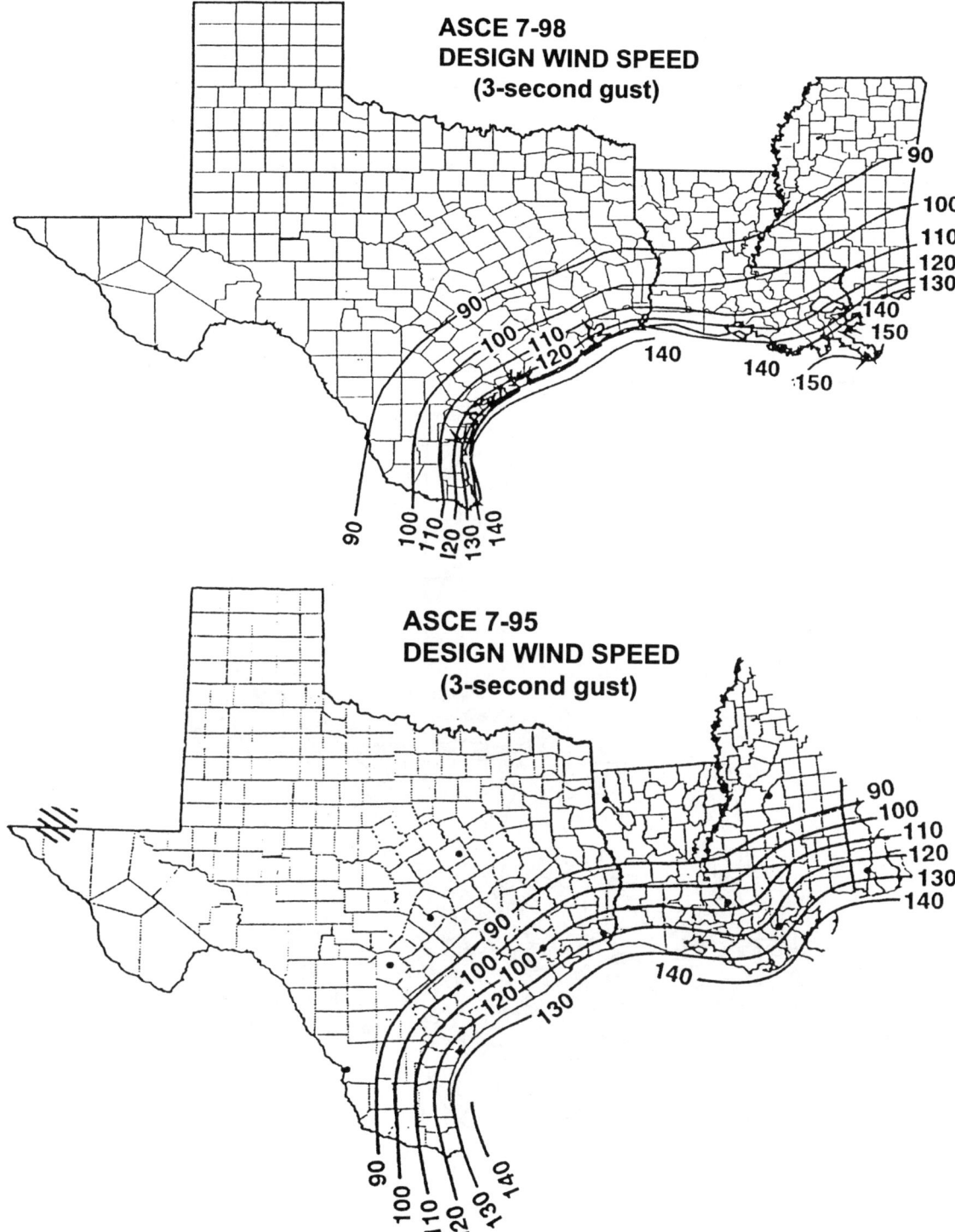

Figure 1.2.1 A Comparison of Basic Design Wind Speeds for 7-98 and 7-95 for Texas, Louisiana, and Mississippi

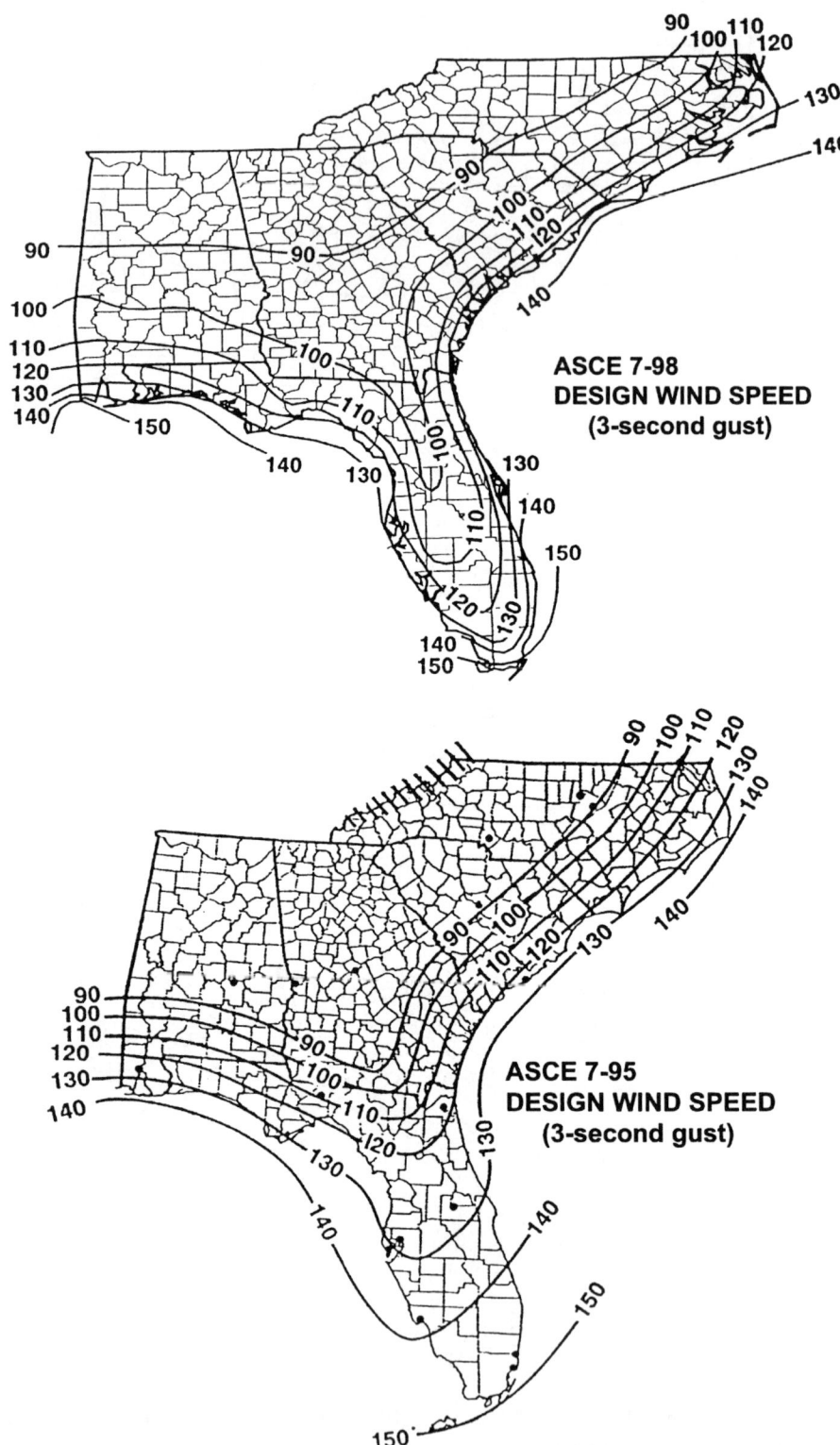

Figure 1.2.2 A Comparison of Basic Design Wind Speeds for 7-98 and 7-95 for Alabama, Florida, Georgia, South Carolina, and North Carolina

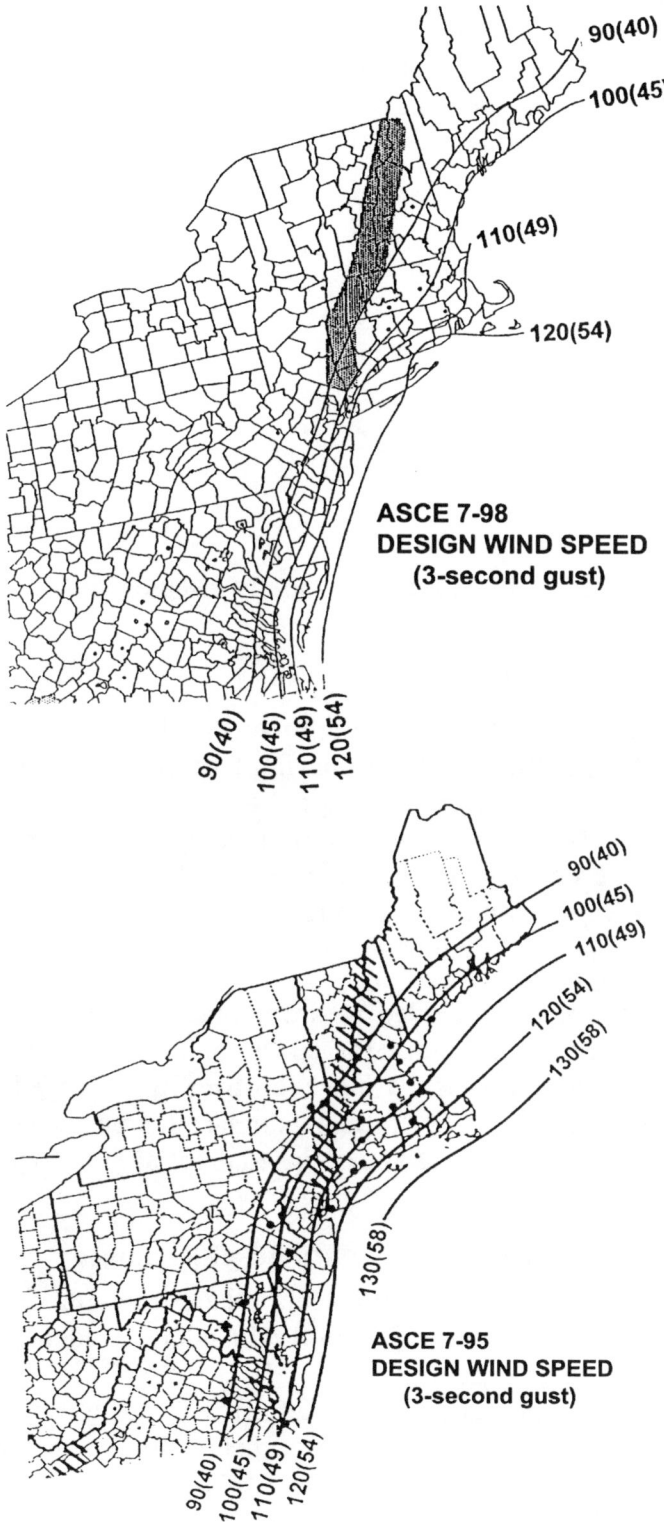

Figure 1.2.3 A Comparison of Basic Design Wind Speeds for 7-98 and 7-95 for Virginia to Maine

1.3   LIMITATIONS OF STANDARD

The possible shortcomings or limitations of the Standard are directly dependent on accurate knowledge of parameters and factors utilized in the algorithms defining the wind loads for design applications:

- Basic wind speed V
- Exposure factor K_z
- Topographic effect K_{zt}
- Directionality factor K_d
- Gust effect factor G
- Pressure and force coefficients
  - C_p, (GC_{pf}) for MWFRS
  - (GC_p) for C&C
  - (GC_{pi})

Limitations of each as used in the Standard are briefly discussed below.

1.3.1   Assessment of Wind Climate

The current Standard provides a more realistic description of wind speeds than previous editions. Perhaps the most serious limitation is that design speeds are not referenced to direction, and potential wind speed anomalies are defined only in terms of special wind regions. These regions include mountain ranges, gorges, or river valleys. Unusual winds may be encountered in these regions because of orographic effects or because of the channeling of wind. The Standard does permit climatological studies using regional climatic data and consultation with a wind engineer and/or a meteorologist.

Tornado winds are not included in development of the basic wind speed map because of their rare occurrence. Intense tornadoes can have ground level wind speeds in the range of 150-200 mph; however, the annual probability of exceedance of this range of wind speeds may be less than 1×10^{-5} (mean recurrence interval exceeding 100,000 years). Special structures and storm shelters can be designed to resist tornado winds if required.

1.3.2   Limitations in Evaluating Structural Response

Given that the majority of buildings and other structures can be treated as rigid structures, the gust effect factor specified in the Standard is adequate. For dynamically sensitive buildings and other structures, a gust effect factor G_f is given. The formulation of gust effect factor G_f is primarily for buildings; it is not always applicable to other structures. It should be noted that the gust effect factor G_f is based on along-wind buffeting response.

Vortex shedding is almost always present with bluff-shaped cylindrical bodies. It can become a problem when the frequency of shedding is close to, or equal to, the frequency of the first or second transverse modes of the structure. The intensity of excitation increases with aspect ratio (height-to-width or length-to-breadth) and decreases with increasing structural damping. Structures with low damping and with aspect ratio of 8 or more may be prone to damaging vortex excitation. If across-wind or torsional excitation appears to be a possibility, expert advice should be obtained.

Another limitation with respect to evaluating structural response is that the Standard fails to define acceptable design wind speeds for serviceability states (deflection, dynamic sway). Table C6-3 in the Commentary does provide conversion factors for determining appropriate wind speeds for mean recurrence intervals of five to 500 years.

1.3.3 Limitations in Shapes of Buildings and Other Structures

The pressure and force coefficients given in the Standard are limited. Many of the structural shapes (e.g., "Y," "T," and "L" shapes) or buildings with stepped elevations are not included. Fortunately, this information may be found in other sources, see Table 1.4.1.

The designer is encouraged to use values available in the literature when coefficients for a specific shape are not given in the Standard. However, the use of prudent judgment is advised, and the following caveats must be addressed:

1. Were the coefficients obtained from proper turbulent boundary layer wind tunnel tests (BLWT) or were they generated under conditions of relatively smooth flow?

2. The averaging time used must necessarily be considered in order to determine whether the coefficients are directly applicable to the evaluation of design loads or they need to be modified.

3. The reference wind speed (fastest-mile, hourly mean, 10-minute mean, 3-second gust, etc.) and exposure category under which the data are generated must be established in order to properly compute the velocity pressure q.

4. If an envelope approach is used, the coefficients should be appropriate for all wind directions. If, however, a directional approach is indicated, then the applicability of the coefficients as a function of wind direction needs to be ascertained. A major limitation in the use of directional coefficients is that their adequacy for other than normal wind directions may not have been verified.

1.4 TECHNICAL LITERATURE

There has been a vast amount of literature published on wind engineering during the past three decades. Most of it is in the form of research papers in the Journal of Wind Engineering and Industrial Aerodynamics, Journal of Structural Engineering, Journal of the American Society of Civil Engineers, Proceedings of the International Conferences on Wind Engineering (a total of 10), Proceedings of the U.S. National Conferences on Wind Engineering (a total of eight), Proceedings of the U.S. Asia-Pacific Conferences on Wind Engineering (a total of four) and Proceedings of the European-African Conferences on Wind Engineering (two). The literature is extensive and scholarly; however, it is not always in a format that can be used by practicing professionals.

Several textbooks, handbooks, standards and codes, reports, and papers contain material that can be used to determine wind loads. Selected items are identified in Table 1.4.1. The items are listed by subject matter for easy identification. Detailed references for theses items are given in the References Section citations.

Table 1.4.1 Technical Literature

Subjects	Selected Reference Material
Wind Effects on Buildings and Structures	Newberry and Eaton (1974); Lawson, vols. 1 and 2 (1980); Cook, vols. 1 and 2 (1985); Holmes, Melbourne, and Walker (1990); Liu (1991); Simiu and Scanlan (1996); Holmes (2001)
Foreign Codes and Standards	NRCC (1995a, 1995b); British Standard BS6399 (1995); Eurocode Draft (1994); ISO (1997); Australian /New Zealand Standard AS/NZS 1170.2 (2001)
Wind Tunnel Testing	Reinhold (1982); ASCE (1987)
General Wind Research	ASCE (1961); Cermak (1977); Davenport (1979, 1978); Simiu (1981)
Pressure and Force Coefficients	ASCE (1961); Hoerner (1965); ASCE (1997)
Tornadoes, Shelter Design	Minor, McDonald, and Mehta (1977); FEMA 83-A (1980); Minor (1982); McDonald (1983); FEMA 320 (1999); FEMA 361 (2000)
Impact Resistance Protocol	SBCCI (1999); ASTM E1886-97, E1996-01; Miami/Dade County Building Code Compliance Office Protocol PA 201-94 and PA 203-94

Chapter 2.
WIND LOAD PROVISIONS

2.1 FORMAT

The wind load provisions of the Standard were completely reformatted to make them more designer friendly and consistent with the formats used by the model code organizations. To assist the designer and the appropriate regulatory agencies in interpreting (and enforcing) specific provisions, many new definitions have been added, see Section 6.2.

The designer is given three options for evaluating the design wind loads for buildings and other structures:

1. Method 1 – Simplified Procedure as specified in Section 6.4 for buildings meeting certain requirements (essentially diaphragm buildings having roof height of 30 ft or less and slopes less than 10 degrees).

2. Method 2 – Analytical Procedure of Section 6.5 applicable to buildings and other structures.

3. Wind Tunnel Procedure as set forth in Section 6.6.

The simplified and analytical procedures provide specific steps to be followed, see Sections 6.4.2 and 6.5.3, respectively, in the determination of wind loads on main wind-force resisting systems (MWFRS) and components and cladding (C&C) separately. MWFRS is defined in Section 6.2 as the overall structure receiving wind loading from more than one surface. Cladding receives wind loads directly and generally transfers the load to other components or to the MWFRS. Equations for the determination of wind loads using the analytical procedures are given in the body of the text. Some of the important footnotes previously given with figures and tables have been moved into the body of the text.

Equations for the graphs of Figures 6-5 through 6-8 in the Standard are given in 2.4 of this document since these graphs as presented in the Standard are difficult for interpolation. In order to avoid repeating the words ASCE 7-98 or the Standard, the section, table, and figure numbers are given without any additional reference.

2.2   DESIGN PROCEDURES

2.2.1   Velocity Pressure

The first step in using Method 2, Analytical Procedure, is to determine the appropriate parameters for evaluating the velocity pressure q.

Velocity pressure, q, at any height above ground and at mean roof height is obtained by the equation:

$$q_z = 0.00256 K_z K_{zt} K_d V^2 I \qquad (lb/ft^2) \qquad (6\text{-}13)$$

where q: effective velocity pressure to be used in the appropriate equations to evaluate wind pressures for MWFRS and C&C; q_z at any height z above ground; q_h is based on K_h at mean roof height h.

K_z: exposure velocity pressure coefficient, which reflects change in wind speed with height and terrain roughness, see Table 6-5.

K_{zt}:   topographic factor which accounts for wind speed-up over hills and escarpments, see Figure 6-2 and $K_{zt} = (1 + K_1 K_2 K_3)^2$.

K_d: directionality factor, see Table 6-6.

V: basic wind speed, which is the 3-second gust speed at 33 ft above ground for Exposure Category C and is approximately associated with an annual probability of 0.02 (50-year mean recurrence interval), see Figure 6-1.

I: importance factor, which adjusts wind speed associated with annual probability of 0.02 (50-year mean recurrence interval) to other probabilities (25-year or 100-year MRI), see Table 6-2.

2.2.2   Method 1 – Simplified Procedure

This method has been added to the Standard to simplify the evaluation of design loads for common regular shape, simple diaphragm buildings having a mean roof height less than or equal to 30 ft and a roof slope less than 10 degrees.

Two tables are provided: Table 6-2 for MWFRS and Tables 6-3A and 6-3B for C&C. For MWFRS, Method 1 combines the windward and leeward pressures into a net horizontal wind pressure on the walls (internal pressures cancel). The maximum uplift on the roof for MWFRS is based on a positive internal pressure as the controlling case. For component and cladding, values are provided for both enclosed buildings and partially enclosed buildings and represent the net pressure (sum of external and internal pressures) applied normal to surfaces. The following values have been assumed in the preparation of the tables:

H = 30 ft;
Exposure B, K_z = 0.70;

$K_d = 0.85$;
$G = 0.85$;
$K_{zt} = 1.0$
$I = 1.0$;
$GC_{pi} = \pm 0.18$ (enclosed building);
$GC_{pi} = \pm 0.55$ (partially enclosed building);
MWFRS pressure coefficients from Figure 6-3;
C&C pressure coefficients from Figure 6-5.

For exposures other than Exposure B, multiplying factors are given. For importance factors other than $I = 1.0$, table values should be multiplied by I.

2.2.3 Method 2 – Analytical Procedure

The analytical procedure for this method is applicable to

1. Buildings of all heights.
2. Low-rise buildings having a mean roof height less than or equal to 60 ft and a gable roof.
3. Open buildings and other structures.

The design procedure for each is given by:

1. The basic wind speed V and wind directionality factor K_d shall be determined in accordance with Section 6.5.4.
2. An importance factor I shall be determined in accordance with Section 6.5.5.
3. An exposure category or exposure categories and velocity pressure exposure coefficient K_z or K_h, as applicable, shall be determined for each wind direction in accordance with Section 6.5.6.
4. A topographic factor K_{zt} shall be determined in accordance with Section 6.5.7.
5. A gust effect factor G or G_f, as applicable, shall be determined in accordance with Section 6.5.8.
6. An enclosure classification shall be determined in accordance with Section 6.5.9.
7. Internal pressure coefficient GC_{pi} shall be determined in accordance with Section 6.5.11.1.
8. External pressure coefficients C_p or GC_{pf}, or force coefficients C_f, as applicable, shall be determined in accordance with Section 6.5.11.2 or 6.5.11.3, respectively.
9. Velocity pressure q_z or q_h, as applicable, shall be determined in accordance with Section 6.5.10.
10. Design wind pressure p or force F shall be determined in accordance with Sections 6.5.12 and 6.5.13, as applicable.

Design pressures for MWFRS and for C&C are determined separately. Generally, C&C design pressures will be higher because of localized high pressures acting over small areas. MWFRS receive wind pressures from several surfaces; hence, with spatial averaging; the pressures are likely to be smaller than those for C&C.

Calculation of design pressures requires selection of appropriate gust effect factors and pressure or force coefficients. The equation for the evaluation of wind loads guide the user in the selection of appropriate factors and coefficients. Various gust effect factors and pressure and force coefficients specified in the Standard are as follows:

G: gust effect factor for MWFRS of buildings (all heights) and for other structures, Section 6.5.8.1

G_f: gust effect factor for MWFRS of flexible buildings and dynamically sensitive other structures obtained using a rational analysis, Section 6.5.8.2.

C_p: external pressure coefficients for MWFRS of buildings (all heights), Figure 6-3.

C_f: force coefficients for open buildings and other structures, Tables 6-9 through 6-13.

(GC_{pf}): external pressure coefficients for MWFRS of low-rise buildings, Figure 6-4.

(GC_p): external pressure coefficients for C&C of buildings, Figures 6-5 through 6-8.

(GC_{pi}): internal pressure coefficients for MWFRS and C&C of buildings, Table 6-7.

Sign convention in the Standard is as follows:

+ (plus sign) means pressure acting toward the surface

- (minus sign) means pressure acting away from the surface

Whenever the sign of ± is specified, both positive and negative values should be used to obtain design loads. Values of external and internal pressures are to be combined algebraically to obtain the most critical load.

2.2.3.1 Design Pressures For MWFRS: Buildings of All Heights

Design wind pressures for the main wind-force resisting systems shall be determined from the following equation:

$$p = qGC_p - q_i(GC_{pi}) \quad (\text{lb/ft}^2)\,(\text{N/m}^2) \qquad \text{(Equation 6-15)}$$

where

q: q_z or q_h as specified in Figure 6-3.
G: gust effect factor given in Section 6.5.8.

C_p: external pressure coefficients given in Figure 6-3.
(GC_{pi}): internal pressure coefficients given in Table 6-7.

2.2.3.2 Design Pressures for MWFRS: Low-Rise Buildings

A low-rise building is a building with mean roof height less than or equal to 60 ft, a gable roof, and with ratio of mean roof height to least horizontal dimension of less than unity. Design pressures for main wind-force resisting systems can be determined from the following equation:

$$p = q_h[(GC_{pf}) - (GC_{pi})] \qquad (\text{lb/ft}^2)(\text{N/m}^2) \qquad \text{(Equation 6-16)}$$

where q_h: velocity pressure at mean roof height
      (GC_{pf}): external pressure coefficients given in Figure 6-4; both loading Cases A and B at each building corner must be considered for the design of framing, see Example 8 in Chapter 3.
      (GC_{pi}): internal pressure coefficients given in Table 6-7.

2.2.3.3 Design Pressures for MWFRS: Flexible Buildings

Design wind pressures for the main wind-force resisting system of flexible buildings shall be determined from the following equation:

$$p = q G_f C_p - q_i(GC_{pi}) \qquad (\text{lb/ft}^2) \; (\text{N/m}^2) \qquad \text{(Equation 6-17)}$$

where q, q_i, C_p and (GC_{pi}) are as defined in Section 6.5.12.2.1, and G_f = gust effect factor defined in Section 6.5.8.2.

2.2.3.4 Design Pressures for C&C: Low-Rise Buildings and All Buildings with h ≤ 60 ft (18.3 m)

Design wind pressures on component and cladding elements of low-rise buildings and buildings with h ≤ 60 ft (18.3 m) shall be determined from the following equation:

$$p = q_h[(GC_p) - (GC_{pi})] \quad (\text{lb/ft}^2) \; (\text{N/m}^2) \qquad \text{(Equation 6-18)}$$

where

    q_h = velocity pressure evaluated at mean roof height h using exposure defined in Section 6.5.6.3.1;
    (GC_p) = external pressure coefficients given in Figures 6-5 through 6-7; and
    (GC_{pi}) = internal pressure coefficient given in Table 6-7.

2.2.3.5 Design Pressures for C&C: Building with h > 60 ft (18.3 m)

Design wind pressures on components and cladding for all buildings with h > 60 ft (18.3 m) shall be determined from the following equation:

$$p = q(GC_p) - q_i(GC_{pi}) \ (lb/ft^2) \ (N/m^2) \qquad \text{(Equation 6-19)}$$

where

q = q_z for windward walls calculated at height z above the ground;

q = q_h for leeward walls, side walls, and roofs evaluated at height h;

q_i = q_h for windward walls, side walls, leeward walls, and roofs of enclosed buildings and for negative internal pressure evaluation in partially enclosed buildings; and

q_i = q_z for positive internal pressure evaluation in partially enclosed buildings where height z is defined as the level of the highest opening in the building that could affect the positive internal pressure. For buildings sited in wind-borne debris regions with glazing in the lower 60 ft (18.3 m) that is not impact resistant or protected with an impact resistant covering, the glazing shall be treated as an opening in accordance with Section 6.5.9.3. For positive internal pressure evaluation, q_i may conservatively be evaluated at height h ($q_i = q_h$);

(GC_p) = external pressure coefficient from Figure 6-8; and

(GC_{pi}) = internal pressure coefficient given in Table 6-7.

q and q_i shall be evaluated using exposure defined in Section 6.5.6.3.2.

2.2.3.6 Alternative Design Wind Pressures for C&C in Buildings with 60 ft (18.3 m) < h < 90 ft (27.4 m)

Alternative to the requirements of Section 2.2.3.5 the design of components and cladding for buildings with a mean roof height greater than 60 ft (18.3 m) and less than 90 ft (27.4 m) values from Figures 6-5, 6-6, and 6-7 shall be used only if the height-to-width ratio is 1 or less, except as permitted by Note 6 of Figure 6-8, and Equation 6-18 is used.

2.2.3.7 Special Provision for Design Pressures for C&C with Tributary Area > 700 sq ft (65 sq m)

Component and cladding elements with tributary area (not wind effective area) greater than 700 sq ft (65 sq m) can be designed using provisions for main wind-force resisting system, see 6.5.12.1.3 of the Standard.

2.2.3.8 Design Wind Loads on Open Buildings and Other Structures

The design wind-force for open buildings and other structures shall be determined by the following formula:

$$F = q_z G C_f A_f \quad \text{(lb)(N)} \qquad \text{(Equation 6-20)}$$

where

q_z = velocity pressure evaluated at height z of the centroid of area A_f using exposure defined in Section 6.5.6.3.2;

G = gust effect factor from Section 6.5.8;

C_f = net force coefficients from Tables 6-9 through 6-12; and

A_f = projected area normal to the wind except where C_f is specified for the actual surface area, ft^2 (m^2).

2.3   WIND TUNNEL PROCEDURE

For those situations where the analytical procedure is considered to be uncertain or inadequate, or where more accurate wind pressures are desired, consideration should be given to wind tunnel tests. The Standard lists a set of conditions in Section 6.6 that must be satisfied for the proper conduct of such tests. The wind tunnel is particularly useful for obtaining detailed information about pressure distributions on complex shapes and the dynamic response of structures. Model scales for structural applications can range from 1:50 for a single-family dwelling to 1:400 for tall buildings. Even smaller scales may be used to model long-span bridges. Of equal importance is the ability to model complex topography at scales of the order of 1:10,000 and assess the effects of features such as hills, mountains, or river gorges on the near-surface winds. Details on wind tunnel modeling for structural or civil engineering applications may be found in ASCE (1987), Cermak (1977), and Reinhold (1982).

2.4   EQUATIONS FOR GRAPHS

Figures 6-5 through 6-8 give external pressure coefficient (GC_p) values for C&C for buildings as a function of effective area of each component and cladding. Wind tunnel results found this relationship between pressure coefficients and effective area to be a logarithmic function. The scale of effective area in the figures is log scale, which makes it very difficult to interpolate. Equations for each of the lines in these figures are given below. The equations can be used to determine wind loads.

Table 2.4.1 Walls for Buildings with h ≤ 60 ft

Positive: Zones 4 and 5, Figure 6-5a

$(GC_p) = 1.0$	for A ≤ 10 sq ft
$(GC_p) = 1.1766 - 0.1766 \log A$	for $10 < A ≤ 500$ sq ft
$(GC_p) = 0.7$	for A > 500 sq ft

Negative: Zone 4

$(GC_p) = -1.1$	for A ≤ 10 sq ft
$(GC_p) = -1.2766 + 0.1766 \log A$	for $10 < A ≤ 500$ sq ft
$(GC_p) = -0.8$	for A > 500 sq ft

Negative: Zone 5

$(GC_p) = -1.4$	for A ≤ 10 sq ft
$(GC_p) = -1.7532 + 0.3532 \log A$	for $10 < A ≤ 500$ sq ft
$(GC_p) = -0.8$	for A > 500 sq ft

Note: Zones are shown in the figures referenced in *ASCE 7-98*.

Table 2.4.2 Gabled Roofs with h ≤ 60 ft

Roofs θ ≤ 10°, Figure 6-5b

Positive with and without overhang: Zones 1, 2, and 3

$(GC_p) = 0.3$ for A ≤ 10 sq ft
$(GC_p) = 0.4000 - 0.1000 \log A$ for 10 < A ≤ 100 sq ft
$(GC_p) = 0.2$ for A > 100 sq ft

Negative without overhang: Zone 1

$(GC_p) = -1.0$ for A ≤ 10 sq ft
$(GC_p) = -1.1000 + 0.1000 \log A$ for 10 < A ≤ 100 sq ft
$(GC_p) = -0.9$ for A > 100 sq ft

Negative without overhang: Zone 2

$(GC_p) = -1.8$ for A ≤ 10 sq ft
$(GC_p) = -2.5000 + 0.7000 \log A$ for 10 < A ≤ 100 sq ft
$(GC_p) = -1.1$ for A > 100 sq ft

Negative without overhang: Zone 3

$(GC_p) = -2.8$ for A ≤ 10 sq ft
$(GC_p) = -4.5000 + 1.7000 \log A$ for 10 < A ≤ 100 sq ft
$(GC_p) = -1.1$ for A > 100 sq ft

Negative with overhang: Zones 1 and 2

$(GC_p) = -1.7$ for A ≤ 10 sq ft
$(GC_p) = -1.8000 + 0.1000 \log A$ for 10 < A ≤ 100 sq ft
$(GC_p) = -3.0307 + 0.7153 \log A$ for 100 < A ≤ 500 sq ft
$(GC_p) = -1.1$ for A > 500 sq ft

Negative with overhang: Zone 3

$(GC_p) = -2.8$ for A ≤ 10 sq ft
$(GC_p) = -4.8000 + 2.0000 \log A$ for 10 < A ≤ 100 sq ft
$(GC_p) = -0.8$ for A > 100 sq ft

Note: Zones are shown in the figures referenced in *ASCE 7-98*.

Table 2.4.3 Gabled and Hipped Roofs with h ≤ 60 ft

Roofs $10° < \theta \leq 30°$, Figure 6-5b

Positive with and without overhang: Zones 1, 2, and 3

$(GC_p) = 0.5$ for $A \leq 10$ sq ft

$(GC_p) = 0.7000 - 0.2000 \log A$ for $10 < A \leq 100$ sq ft

$(GC_p) = 0.3$ for $A > 100$ sq ft

Negative with and without overhang: Zone 1

$(GC_p) = -0.9$ for $A \leq 10$ sq ft

$(GC_p) = -1.0000 + 0.1000 \log A$ for $10 < A \leq 100$ sq ft

$(GC_p) = -0.8$ for $A > 100$ sq ft

Negative without overhang: Zones 2 and 3

$(GC_p) = -2.1$ for $A \leq 10$ sq ft

$(GC_p) = -2.8000 + 0.7000 \log A$ for $10 < A \leq 100$ sq ft

$(GC_p) = -1.4$ for $A > 100$ sq ft

Negative with overhang: Zone 2

$(GC_p) = -2.2$ for all A sq ft

Negative with overhang: Zone 3

$(GC_p) = -3.7$ for $A \leq 10$ sq ft

$(GC_p) = -4.9000 + 1.2000 \log A$ for $10 < A \leq 100$ sq ft

$(GC_p) = -2.5$ for $A > 100$ sq ft

Note: Zones are shown in the figures referenced in *ASCE 7-98*.

Table 2.4.4 Gabled Roofs with h ≤ 60 ft

Roofs 30° < θ ≤ 45°, Figure 6-5b

Positive with and without overhang: Zones 1, 2, and 3

$(GC_p) = 0.9$	for $A \leq 10$ sq ft
$(GC_p) = 1.0000 - 0.1000 \log A$	for $10 < A \leq 100$ sq ft
$(GC_p) = 0.8$	for $A > 100$ sq ft

Negative with and without overhang: Zone 1

$(GC_p) = -1.0$	for $A \leq 10$ sq ft
$(GC_p) = -1.2000 + 0.2000 \log A$	for $10 < A \leq 100$ sq ft
$(GC_p) = -0.8$	for $A > 100$ sq ft

Negative without overhang: Zones 2 and 3

$(GC_p) = -1.2$	for $A \leq 10$ sq ft
$(GC_p) = -1.4000 + 0.2000 \log A$	for $10 < A \leq 100$ sq ft
$(GC_p) = -1.0$	for $A > 100$ sq ft

Negative with overhang: Zones 2 and 3

$(GC_p) = -2.0$	for $A \leq 10$ sq ft
$(GC_p) = -2.2000 + 0.2000 \log A$	for $10 < A \leq 100$ sq ft
$(GC_p) = -1.8$	for $A > 100$ sq ft

Note: Zones are shown in the figures referenced in *ASCE 7-98*.

Table 2.4.5 Multispan Gabled Roofs with h ≤ 60 ft

Roofs 10° < θ ≤ 30°, Figure 6-6

Positive: Zones 1, 2, and 3

$(GC_p) = 0.6$ for $A \leq 10$ sq ft
$(GC_p) = 0.8000 - 0.2000 \log A$ for $10 < A \leq 100$ sq ft
$(GC_p) = 0.4$ for $A > 100$ sq ft

Negative: Zone 1

$(GC_p) = -1.6$ for $A \leq 10$ sq ft
$(GC_p) = -1.8000 + 0.2000 \log A$ for $10 < A \leq 100$ sq ft
$(GC_p) = -1.4$ for $A > 100$ sq ft

Negative: Zone 2

$(GC_p) = -2.2$ for $A \leq 10$ sq ft
$(GC_p) = -2.7000 + 0.5000 \log A$ for $10 < A \leq 100$ sq ft
$(GC_p) = -1.7$ for $A > 100$ sq ft

Negative: Zone 3

$(GC_p) = -2.7$ for $A \leq 10$ sq ft
$(GC_p) = -3.7000 + 1.0000 \log A$ for $10 < A \leq 100$ sq ft
$(GC_p) = -1.7$ for $A > 100$ sq ft

Note: Zones are shown in the figures referenced in *ASCE 7-98*.

Table 2.4.6 Multispan Gabled Roofs with h ≤ 60 ft

Roofs 30° < θ ≤ 45°, Figure 6-6

Positive: Zones 1, 2, and 3

$(GC_p) = 1.0$	for A ≤ 10 sq ft
$(GC_p) = 1.2000 - 0.2000 \log A$	for 10 < A ≤ 100 sq ft
$(GC_p) = 0.8$	for A > 100 sq ft

Negative: Zone 1

$(GC_p) = -2.0$	for A ≤ 10 sq ft
$(GC_p) = -2.9000 + 0.9000 \log A$	for 10 < A ≤ 100 sq ft
$(GC_p) = -1.1$	for A > 100 sq ft

Negative: Zone 2

$(GC_p) = -2.5$	for A ≤ 10 sq ft
$(GC_p) = -3.3000 + 0.8000 \log A$	for 10 < A ≤ 100 sq ft
$(GC_p) = -1.7$	for A > 100 sq ft

Negative: Zone 3

$(GC_p) = -2.6$	for A ≤ 10 sq ft
$(GC_p) = -3.5000 + 0.9000 \log A$	for 10 < A ≤ 100 sq ft
$(GC_p) = -1.7$	for A > 100 sq ft

Note: Zones are shown in the figures referenced in *ASCE 7-98*.

Table 2.4.7 Monoslope Roofs with h ≤ 60 ft

Roofs 3° < θ ≤ 10°, Figure 6-7a

Positive: All Zones

$(GC_p) = 0.3$	for $A \leq 10$ sq ft
$(GC_p) = 0.4000 - 0.1000 \log A$	for $10 < A \leq 100$ sq ft
$(GC_p) = 0.2$	for $A > 100$ sq ft

Negative: Zone 1

$(GC_p) = -1.1$	for all A sq ft

Negative: Zone 2

$(GC_p) = -1.3$	for $A \leq 10$ sq ft
$(GC_p) = -1.4000 + 0.1000 \log A$	for $10 < A \leq 100$ sq ft
$(GC_p) = -1.2$	for $A > 100$ sq ft

Negative: Zone 2'

$(GC_p) = -1.6$	for $A \leq 10$ sq ft
$(GC_p) = -1.7000 + 0.1000 \log A$	for $10 < A \leq 100$ sq ft
$(GC_p) = -1.5$	for $A > 100$ sq ft

Negative: Zone 3

$(GC_p) = -1.8$	for $A \leq 10$ sq ft
$(GC_p) = -2.4000 + 0.6000 \log A$	for $10 < A \leq 100$ sq ft
$(GC_p) = -1.2$	for $A > 100$ sq ft

Negative: Zone 3'

$(GC_p) - -2.6$	for $A \leq 10$ sq ft
$(GC_p) = -3.6000 + 1.0000 \log A$	for $10 < A \leq 100$ sq ft
$(GC_p) = -1.6$	for $A > 100$ sq ft

Note: Zones are shown in the figures referenced in *ASCE 7-98*.

Table 2.4.8 Monoslope Roofs with h ≤ 60 ft

Roofs 10° < θ ≤ 30°, Figure 6-7a

Positive: All Zones

$(GC_p) = 0.4$ for $A \leq 10$ sq ft

$(GC_p) = 0.5000 - 0.1000 \log A$ for $10 < A \leq 100$ sq ft

$(GC_p) = 0.3$ for $A > 100$ sq ft

Negative: Zone 1

$(GC_p) = -1.3$ for $A \leq 10$ sq ft

$(GC_p) = -1.5000 + 0.2000 \log A$ for $10 < A \leq 100$ sq ft

$(GC_p) = -1.1$ for $A > 100$ sq ft

Negative: Zone 2

$(GC_p) = -1.6$ for $A \leq 10$ sq ft

$(GC_p) = -2.0000 + 0.4000 \log A$ for $10 < A \leq 100$ sq ft

$(GC_p) = -1.2$ for $A > 100$ sq ft

Negative: Zone 3

$(GC_p) = -2.9$ for $A \leq 10$ sq ft

$(GC_p) = -3.8000 + 0.9000 \log A$ for $10 < A \leq 100$ sq ft

$(GC_p) = -2.0$ for $A > 100$ sq ft

Note: Zones are shown in the figures referenced in *ASCE 7-98*.

Table 2.4.9 Sawtooth Roofs with $h \leq 60$ ft

Roofs $\theta \geq 10°$, Figure 6-7b

Positive: Zone 1

$\quad (GC_p) = 0.7$ for $A \leq 10$ sq ft

$\quad (GC_p) = 0.8766 - 0.1766 \log A$ for $10 < A \leq 500$ sq ft

$\quad (GC_p) = 0.4$ for $A > 500$ sq ft

Positive: Zone 2

$\quad (GC_p) = 1.1$ for $A \leq 10$ sq ft

$\quad (GC_p) = 1.4000 - 0.3000 \log A$ for $10 < A \leq 100$ sq ft

$\quad (GC_p) = 0.8$ for $A > 100$ sq ft

Positive: Zone 3

$\quad (GC_p) = 0.8$ for $A \leq 10$ sq ft

$\quad (GC_p) = 0.9000 - 0.1000 \log A$ for $10 < A \leq 100$ sq ft

$\quad (GC_p) = 0.7$ for $A > 100$ sq ft

Negative: Zone 1

$\quad (GC_p) = -2.2$ for $A \leq 10$ sq ft

$\quad (GC_p) = -2.8474 + 0.6474 \log A$ for $10 < A \leq 500$ sq ft

$\quad (GC_p) = -1.1$ for $A > 500$ sq ft

Negative: Zone 2

$\quad (GC_p) = -3.2$ for $A \leq 10$ sq ft

$\quad (GC_p) = -4.1418 + 0.9418 \log A$ for $10 < A \leq 500$ sq ft

$\quad (GC_p) = -1.6$ for $A > 500$ sq ft

Negative: Zone 3 (span A)

$\quad (GC_p) = -4.1$ for $A \leq 10$ sq ft

$\quad (GC_p) = -4.5000 + 0.4000 \log A$ for $10 < A \leq 100$ sq ft

$\quad (GC_p) = -8.2782 + 2.2891 \log A$ for $100 < A \leq 500$ sq ft

$\quad (GC_p) = -2.1$ for $A > 500$ sq ft

Negative: Zone 3 (spans B, C, D)

$\quad (GC_p) = -2.6$ for $A \leq 100$ sq ft

$\quad (GC_p) = -4.6030 + 1.0015 \log A$ for $100 < A \leq 500$ sq ft

$\quad (GC_p) = -1.9$ for $A > 500$ sq ft

Note: Zones are shown in the figures referenced in *ASCE 7-98*.

Table 2.4.10 Roof and Walls for Buildings with h > 60 ft

Roofs $\theta \leq 10°$, Figure 6-8

Negative: Zone 1

$(GC_p) = -1.4$	for $A \leq 10$ sq ft
$(GC_p) = -1.6943 + 0.2943 \log A$	for $10 < A \leq 500$ sq ft
$(GC_p) = -0.9$	for $A > 500$ sq ft

Negative: Zone 2

$(GC_p) = -2.3$	for $A \leq 10$ sq ft
$(GC_p) = -2.7120 + 0.4120 \log A$	for $10 < A \leq 500$ sq ft
$(GC_p) = -1.6$	for $A > 500$ sq ft

Negative: Zone 3

$(GC_p) = -3.2$	for $A \leq 10$ sq ft
$(GC_p) = -3.7297 + 0.5297 \log A$	for $10 < A \leq 500$ sq ft
$(GC_p) = -2.3$	for $A > 500$ sq ft

Walls All θ

Positive: Zones 4 and 5

$(GC_p) = 0.9$	for $A \leq 20$ sq ft
$(GC_p) = 1.1792 - 0.2146 \log A$	for $20 < A \leq 500$ sq ft
$(GC_p) = 0.6$	for $A > 500$ sq ft

Negative: Zone 4

$(GC_p) = -0.9$	for $A \leq 20$ sq ft
$(GC_p) = -1.0861 + 0.1431 \log A$	for $20 < A \leq 500$ sq ft
$(GC_p) = -0.7$	for $A > 500$ sq ft

Negative: Zone 5

$(GC_p) = -1.8$	for $A \leq 20$ sq ft
$(GC_p) = -2.5445 + 0.5723 \log A$	for $20 < A \leq 500$ sq ft
$(GC_p) = -1.0$	for $A > 500$ sq ft

Note: Zones are shown in the figures referenced in *ASCE 7-98*.

Chapter 3.
EXAMPLES

Ten illustrative examples of determination of wind loads using the analytical procedure of the ASCE 7-98 are presented to provide guidance to the user of the Standard. The examples include commercial buildings, a house, a mid-rise office building, and a billboard sign. The examples are as follows:

1. 30 ft x 60 ft x 15 ft commercial building with concrete masonry units (CMU) walls
2. Example 1 using simplified procedure
3. 100 ft x 200 ft x 160 ft high office building
4. Office building of Example 3 located on escarpment
5. 2500 sq ft house with gable/hip roof
6. House of Example 5 on an isolated hill
7. 200 ft x 250 ft gable roof commercial/warehouse building using all height provisions
8. Building of Example 7 using low-rise building provisions
9. 40 ft x 80 ft commercial building with monoslope roof with overhang
10. 50 ft x 20 ft billboard sign on poles (flexible) 60 ft above ground.

These examples represent a variety of situations in determination of wind loads. The equation, table, figure, and section numbers of ASCE 7-98 are cited where appropriate. Every effort has been made to check the accuracy of the numbers, though no absolute assurance can be given.

3.1 Example 1 – 30 ft x 60 ft x 15 ft Commercial Building with Concrete Masonry Units Walls

In this example, design wind pressures for a typical load-bearing one-story masonry building are determined. The building data are as follows:

Location: Corpus Christi, Texas
Topography: Homogenous
Terrain: Flat, open terrain
Dimensions: 30 ft x 60 ft x 15 ft, flat roof
Framing: CMU walls on three sides
Steel framing in front with glass
Open web joists, 30 ft span spaced at 5 ft on center, covered with metal panel to provide roof diaphragm action
Cladding: Roof metal panels are 2 ft wide, 20 ft long
Doors and glass size vary, glass is debris resistant

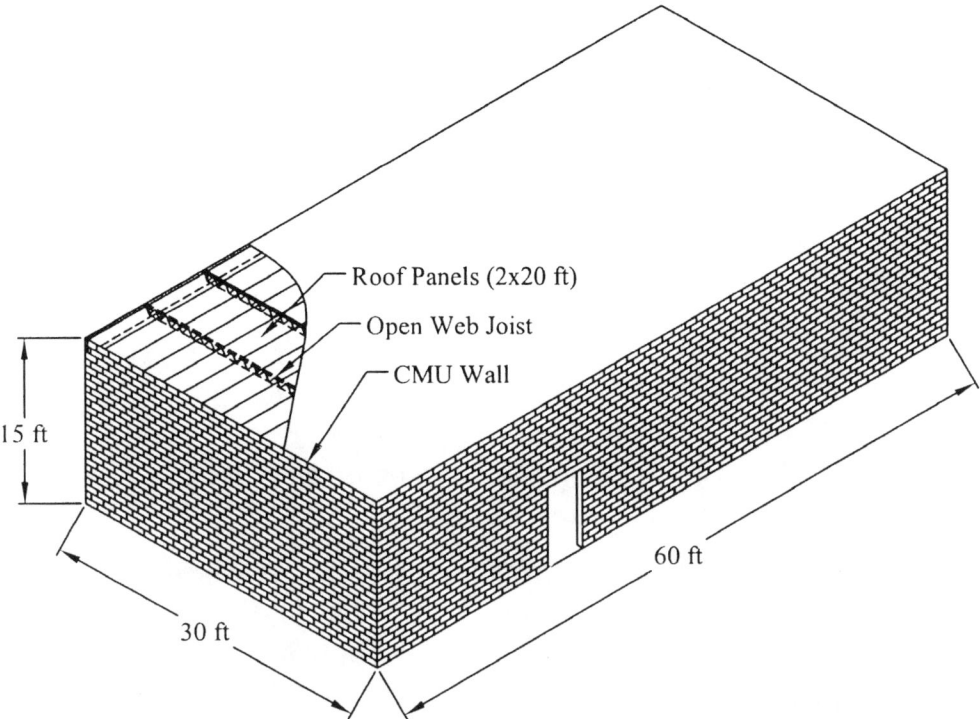

Roof Panels (2x20 ft)
Open Web Joist
CMU Wall

15 ft

60 ft

30 ft

Figure 3.1.1 Building Dimensions for Example 1

For this example, the analytical procedure of ASCE 7-98 for rigid buildings of all heights is used. The same building is illustrated in Example 2 using the simplified procedure of Section 6.4 of ASCE 7-98.

Basic Wind Speed

Selection of the basic wind speed is addressed in Section 6.5.4 of the Standard. Basic wind speed for Corpus Christi, Texas, is 130 mph, see Figure 6-1a.

Exposure

The building is located on flat and open terrain; therefore, use Exposure C Section 6.5.6. Note that Exposure C is valid even if the building is close to the oceanfront.

Building Classification

The building function is shops. It is not considered an essential facility. Building Category II is appropriate.

Velocity Pressure

The velocity pressures are computed using:

$$q_z = 0.00256 \, K_z \, K_{zt} \, K_d \, V^2 \, I \qquad \text{(Equation 6-1)}$$

where: K_z = 0.85 from Table 6-5 for Case 1 (C&C) and Case 2 (MWFRS); for 0-15 ft, there is only one value

K_{zt} = 1.0 for homogeneous topography, see Section 6.5.7
K_d = 0.85 for buildings, see Table 6-6
V = 130 mph, see Figure 6-1a
I = 1.0 for Category II building, see Table 6-1

$$q_z = 0.00256 \ (0.85) \ (1.0) \ (0.85) \ (130)^2 \ (1.0)$$
$$= 31.3 \text{ psf}$$

$$q_h = 31.3 \text{ psf for } h = 15 \text{ ft}$$

Gust Effect Factor

The building is considered a rigid structure. Section 6.5.8.1 permits use of G = 0.85.

If the detailed procedure for a rigid structure is used, the calculated value of G = 0.89; however, the Standard permits the use of one value of G = 0.85. Detailed calculations of G value are illustrated in Example 3.

Use G = 0.85 for this example.

Internal Pressure Coefficient

The building is located in a hurricane-prone area. Section 6.5.9.3 requires that glazing be considered openings unless it is protected or debris resistant.

The example building has debris resistant glazing and other openings are such that it does not qualify for partially enclosed building.

Use (GC_{pi}) = +0.18 and -0.18 for enclosed building, see Table 6-7.

Design Wind Pressures for MWFRS

Design wind pressures are determined using the equation

$$p = q \ G \ C_p - q_i \ (GC_{pi}) \qquad \text{(Equation 6-15)}$$

where: $q = q_z$ for windward wall; it is 31.3 psf for this example
$\quad = q_h$ for leeward wall, side walls, and roof
G = 0.85
C_p = values of external pressure coefficients
$q_i = q_h$ for enclosed building, 31.3 psf
(GC_{pi}) = +0.18 and -0.18

The values of external pressure coefficients are obtained from Figure 6-3.

Wall C_p

The windward wall pressure coefficient is 0.8.

The side wall pressure coefficient is -0.7.

The leeward wall pressure coefficients are a function of L/B ratio:

For L/B = 0.5, C_p = -0.5 for wind normal to 60 ft

For L/B = 2.0, C_p = -0.3 for wind normal to 30 ft

Roof C_p

The roof pressure coefficients are a function of roof slope and h/L. For θ < 10 degrees and h/L = 0.25 and 0.5,

C_p = -0.9 for distance 0 to h

C_p = -0.5 for distance h to 2h

C_p = -0.3 for distance >2h

MWFRS Pressures

Windward wall	p = 31.3 (0.85) (0.8) – 31.3 (±0.18) = 21.3 ± 5.6 psf
Leeward wall	p = 31.3 (0.85) (-0.5) – 31.3 (±0.18) = 13.3 ± 5.6 psf * wind normal to 60 ft
Leeward wall	p = 31.3 (0.85) (-0.3) – 31.3 (±0.18) = -8.0 ± 5.6 psf * wind normal to 30 ft
Roof	p = 31.3 (0.85) (-0.9) – 31.3 (±0.18) = -23.9 ± 5.6 psf for 0 to 15 ft = -13.3 ± 5.6 psf for 15 to 30 ft = -8.0 ± 5.6 psf for > 30 ft

The MWFRS design pressures for two directions are shown in Figures 3.1.2 and 3.1.3. The internal pressures shown are to be added to external pressures as appropriate. The internal pressures of the same sign act on all surfaces; thus, they cancel out for total horizontal shear.

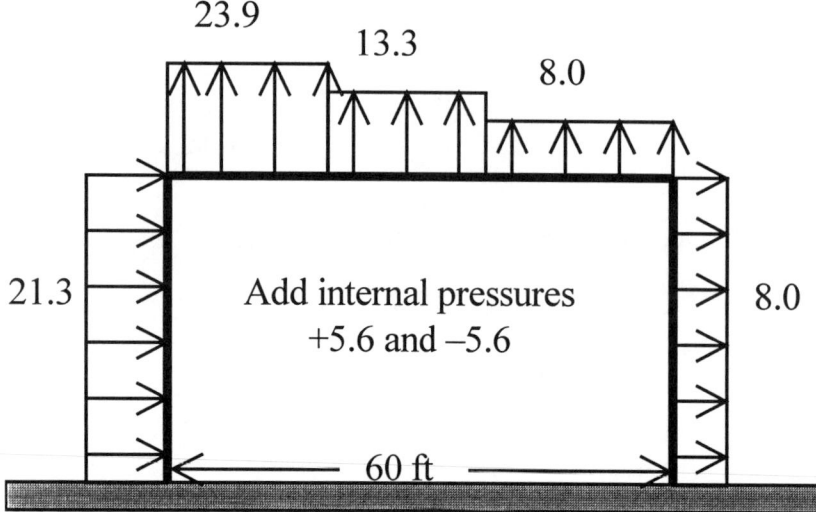

Figure 3.1.2 Design pressures for MWFRS when wind is normal to 30 ft wall.

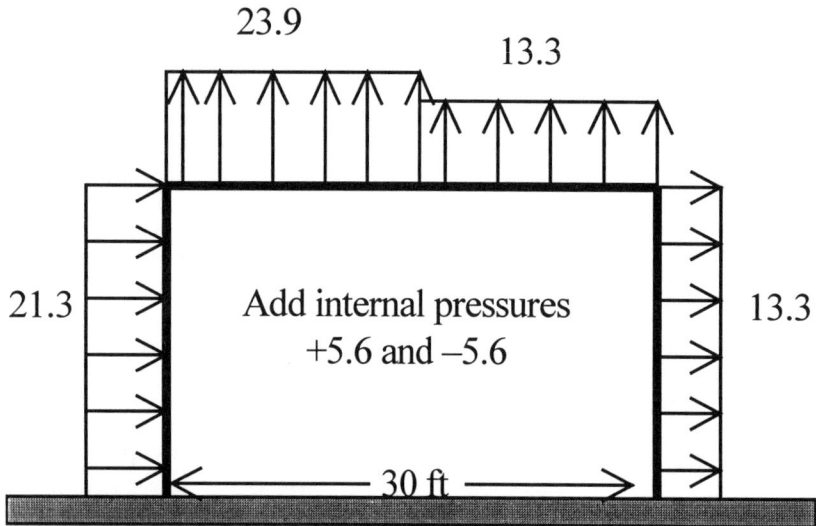

Figure 3.1.3 Design pressures for MWFRS when wind is normal to 60 ft wall.

<u>Design Pressures for C&C</u>

Design wind pressures are determined using the equation

$$p = q_h \left[(GC_p) - (GC_{pi}) \right] \qquad \text{(Equation 6-18)}$$

where: q_h = 31.3 psf

(GC_p) = values obtained from Figure 6-5; they are a function of effective area and zone

$(GC_{pi}) = +0.18$ and -0.18

Wall Pressures

CMU walls are supported at the roof diaphragm and at ground, span = 15 ft.

CMU wall effective wind area is determined using the definition, "the width of effective area need not be less than one-third of the span," see Section 6.2:

CMU wall effective wind area, A = 15 (15/3) = 75 sq ft

In Figure 6-5a, Note 5 suggests that the pressure coefficient values for walls can be reduced by 10% for roof slope of 10 degrees or less. The values of (GC_p) are obtained from the figure or from equations of the graphs, see Section 2.4.

Corner Zone 5 distance:
   smaller of
          a = 0.1 (30) = 3 ft
   or
          = 0.4 (15) = 6 ft
   a = 3 ft controls

Corner Zone 5,   p = 31.3 [(-1.09) (0.9) – (±0.18)] = -36.3 psf

(handwritten: 0.85 above -1.09; +29.7 above -36.3)

                p = 3.13 [(0.85) (0.9) – (±0.18)] = +29.7 psf

(handwritten: 31.3 below 3.13)

Middle Zone 4,   p = 31.3 [(-0.95) (0.9) – (±0.18)] = -32.6 psf
                p = 3.13 [(0.85) (0.9) – (±0.18)] = +29.7 psf

Note: The CMU walls have uplift pressure from the roof, which should be determined on the basis of MWFRS.

Pressure for glazing and mullions can be determined similarly with the known effective wind area.

Roof Joist Pressures

Roof joists span 30 ft and are spaced 5 ft apart. The joist can be in Zone 1 (interior of roof) or Zone 2 (eave area). Zone 3 (roof corner area) acts only on a part of the joist.

Width of Zones 2 and 3:   (Figure 6-5)
   smaller of
          a = 0.1 (30) = 3 ft (controls)
   or
          a = 0.4 (15) = 6 ft

Joist Effective Wind Area:
  larger of
$$A = 30 \times 5 = 150 \text{ ft}$$
  or
$$A = 30 \times (30/3) = 300 \text{ ft (controls)}$$

The values of (GC_p) are obtained from Figure 6-5b or from equations of the graphs using effective area $A = 300$ sq ft.

Interior Zone 1
$$p = 31.3 \, [-0.9 \pm 0.18] = -33.8 \text{ psf}$$
$$p = 31.3 \, [+0.2 \pm 0.18] = +11.9 \text{ psf}$$

Eave Zone 2 and Corner Zone 3 _-40.1_
$$p = 31.3 \, [-0.9 \pm 0.18] = -33.8 \text{ psf}$$
$$p = 31.3 \, [+0.2 \pm 0.18] = +11.9 \text{ psf}$$

Roof Panel Pressures

Even though roof panel length is 20 ft, each panel spans 5 ft between joists.

Roof Panel Effective Area:
  larger of
$$A = 5 \times 2 = 10 \text{ sq ft (controls)}$$
  or
$$A = 5 \times (5/3) = 8 \text{ sq ft}$$
          Width of Zones 2 and 3, $A = 3$ ft

Interior Zone 1
$$p = 31.3 \, [-1.0 \pm 0.18] = -36.9 \text{ psf}$$
$$p = 31.3 \, [+0.3 \pm 0.18] = +15.0 \text{ psf}$$

Eave Zone 2
$$p = 31.3 \, [-1.8 \pm 0.18] = -62.0 \text{ psf}$$
$$p = 31.3 \, [+0.3 \pm 0.18] = +15.0 \text{ psf}$$

Corner Zone 3
$$p = 31.3 \, [-2.8 \pm 0.18] = -93.3 \text{ psf}$$
$$p = 31.3 \, [+0.3 \pm 0.18] = +15.0 \text{ psf}$$

Notes:

- Internal pressure coefficient of +0.18 or -0.18 is used to give critical pressures.

- The roof panel fasteners design pressures will be the same as metal panel since values of (GC_p) are the same for wind effective areas less than 10 sq ft.

3.2   Example 2 – Example 1 Using Simplified Procedure

In this example, design wind pressures for the building of Example 1 are determined using the simplified procedure of Section 6.4. Data for the building are the same as Example 1:

Location: Corpus Christi, Texas
Topography: Homogenous
Terrain: Flat, open terrain
Dimensions: 30 ft x 60 ft x 15 ft, flat roof
Framing: Concrete masonry units (CMU) walls on three sides
                     Steel framing in front with glass
                     Open web joists, 30 ft span spaced at 5 ft on center, covered with
                     metal panel to provide roof diaphragm action
Cladding: Roof metal panels are 2 ft wide, 20 ft long
                     Doors and glass size vary, glass is debris resistant

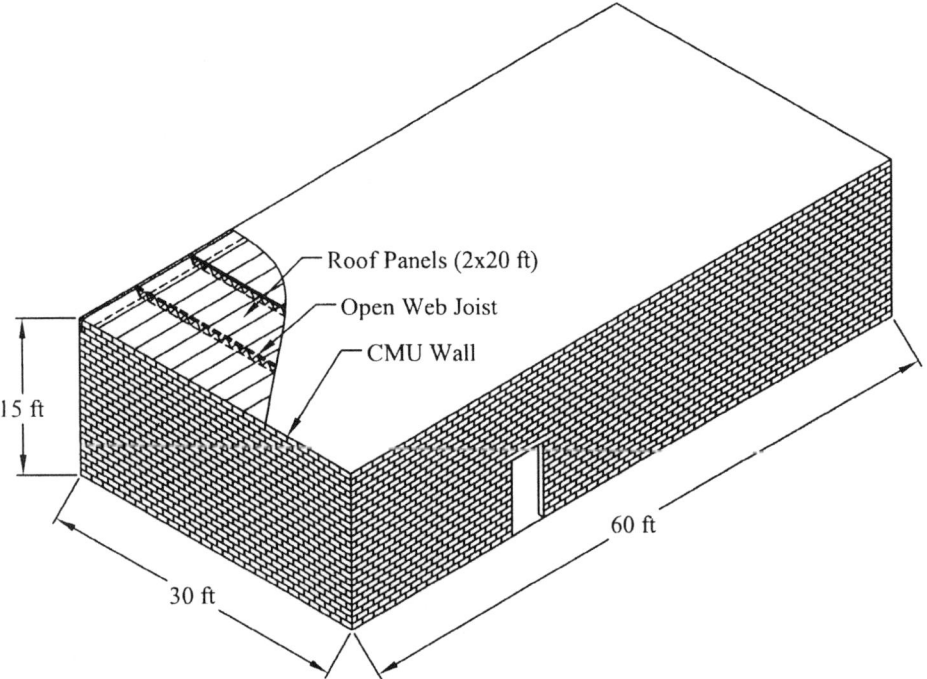

Figure 3.2.1 Building Dimensions for Example 2

In order to use the simplified procedure, all seven conditions of Section 6.4.1 must be satisfied.
1. It is a simple diaphragm building.
2. It has a roof slope of less than 10 degrees.
3. Mean roof height is less than 30 ft.
4. It has a regular shape.
5. It is a rigid building (h/width << 4), see Commentary
6. There is no expansion joint.

7. There is no abrupt change in topography, see Section 6.5.7.1, for requirements of topographic effects.

Basic Wind Speed

Basic wind speed for Corpus Christi, Texas, is 130 mph, see Figure 6-1a.

Exposure

Flat, open terrain constitutes Exposure C, see Section 6.5.6. Note that wind pressure values given in Tables 6-2 and 6-3 are for Exposure B.

Enclosure Classification

Since the building has debris resistant glazing and no dominant opening in any one wall, it can be classified as an enclosed building.

Design Wind Pressures for MWFRS

In Table 6-2, for V = 130 mph and enclosed building,

Roof pressure = -33 psf

Wall pressure = 29 psf

These pressures have to be modified for Exposure C and roof pressure should be multiplied by reduction factor for Exposure C of 1.4, see Note 2, Table 6-2.

As the tributary roof area = 30 x 60 = 1800 sq ft, a reduction factor = 0.8 may be applied to the

Design roof pressure p = (-33) (1.4) (0.8) = -37.0 psf

Design wall pressure p = (29) (1.4) = 40.6 psf

In the simplified procedure, design roof pressure is assumed to act uniformly over the entire roof and it includes internal pressure. The wall pressure is the combined windward and leeward wall pressures (internal pressure cancels).

The analytical procedure used in Example 1 yields roof pressure varying from 13.6 psf to –29.5 psf (including internal pressure) and combined wall pressures of 29.3 psf and 34.6 psf, see Figures 3.1.2 and 3.1.3. The maximum pressures are 15% to 20% lower than the simplified procedure. The uplift values for the roof are even lower where they vary along the roof.

Design Pressures for C&C

In Table 6-3, the values shown are for Exposure B. The values are to be multiplied by a factor of 1.4 for Exposure C, see Note 2 in Table 6-3.

Wall Pressures

The effective wind area for a CMU wall is 75 sq ft, see Example 1. Linear interpolation is permitted in Table 6-3.

From Table 6-3A, for V = 130 mph, for effective wind area of 75 sq ft and for Exposure C, the design pressures are:

Zone 4   p = (+26.8) (1.4) = +37.5 psf
              p = (-29.7) (1.4) = -41.6 psf

Zone 5   p = (+26.8) (1.4) = +37.5 psf
              p = (-33.5) (1.4) = -46.9 psf

Roof Joist Pressures

From Table 6-3A, for V = 130 mph, for effective wind area of 300 sq ft and for Exposure C, the design pressures are:

Zone 1   p = (+10) (1.4) = +14.0 psf
              p = (-28) (1.4) = -39.2 psf

Zones 2 and 3   p = (+10) (1.4) = +14.0 psf
              p = (-33) (1.4) = -46.2 psf

Roof Panel Pressures

Effective wind area for roof panel is 10 sq ft, see Example 1.

From Table 6-3A, for V = 130 mph, for effective wind area of 10 sq ft and for Exposure C, the design pressures are:

Zone 1   p = (+12) (1.4) = +16.8 psf
              p = (-30) (1.4) = -42 psf

Zone 2   p = (+12) (1.4) = +16.8 psf
              p = (-51) (1.4) = -71.4 psf

Zone 3   p = (+12) (1.4) = +16.8 psf
              p = (-77) (1.4) = -107.8 psf

The analytical procedure in Example 1 yields C&C design pressures 12% to 22% lower than the simplified procedure.

3.3   Example 3 – 100 ft x 200 ft x 160 ft High Office Building

Location:	Houston, Texas
Topography:	Flat
Terrain:	Suburban
Dimensions:	100 ft x 200 ft in plan
	Roof height of 157 ft with 3 ft parapet
	Flat roof
Framing:	Reinforced concrete rigid frame in both directions
	Floor and roof slabs provide diaphragm action
	Fundamental natural frequency is greater than 1 Hz
	(Since the height to least horizontal dimension is less than 4, the fundamental frequency is judged to be greater than 1 Hz.)
Cladding:	Mullions for glazing panels span 11 ft between floor slabs
	Mullion spacing is 5 ft
	Glazing panels are 5 ft wide x 5 ft 6 in. high (typical); they are resistant to wind-borne debris impact in the bottom 60 ft.

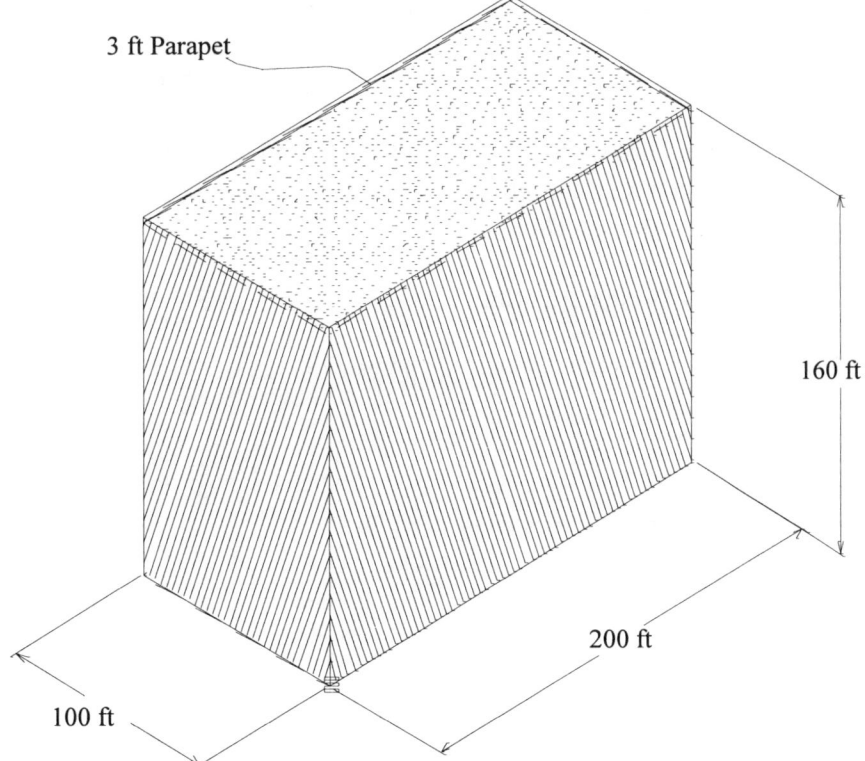

3 ft Parapet

160 ft

200 ft

100 ft

Figure 3.3.1 100 ft x 200 ft x 160 ft Building

Analytical procedure of ASCE 7-98 is to be used.

Exposure

The building is located in a suburban area; therefore, use Exposure B, see Section 6.5.6.

Building Classification

The building function is office space. It is not considered an essential facility or likely to be occupied by 300 persons in a single area at one time. Therefore, building Category II is appropriate, see Table 1-1.

Basic Wind Speed

Selection of the basic wind speed is addressed in Section 6.5.4 of the Standard. Houston, Texas, is located on the 120 mph contour. The basic wind speed V = 120 mph, see Figure 6-1a.

Velocity Pressures

The velocity pressures are computed using the following equation:

$$q_z = 0.00256 K_z K_{zt} K_d V^2 I \text{ psf} \qquad \text{(Equation 6-1)}$$

where: K_z is obtained from Table 6-5, Case 1 for C&C and Case 2 for MWFRS
$K_{zt} = 1.0$ for homogeneous topography
$K_d = 0.85$ for buildings, see Table 6-6
V = 120 mph
I = 1.0 for Category II classification, see Table 6-1

$$q_z = 0.00256 K_z (1.0)(0.85)(120)^2(1.0)$$
$$q_z = 31.3 K_z \text{ psf}$$

Values for K_z and the resulting velocity pressures are given in Table 3.5.1. The velocity pressure at mean roof height, q_h, is 35.1 psf.

Table 3.3.1 q_z Velocity pressures

Height, ft	MWFRS		C&C	
	K_z	q_z, psf	K_z	q_z, psf
0-15	0.57	17.9	0.70	21.9
30	0.70	21.9	0.70	21.9
50	0.81	25.4	0.81	25.4
80	0.93	29.1	0.93	29.1
120	1.04	32.6	1.04	32.6
Roof ht = 157	1.12	35.1	1.12	35.1
Parapet ht = 160	1.13	35.4	1.13	35.4

Design Wind Pressures for the Main Wind-Force Resisting System (MWFRS)

Equations for the design pressures for a building is given by Equation 6-15 of the Standard:

$$p = qGC_p - q_i(GC_{pi})$$

where: $q = q_z$ for windward wall at height z above ground

$q = q_h$ for leeward wall, side walls, and roof

$q_i = q_h$ for enclosed building (see internal pressure coefficient)

G = gust effect factor for rigid building and structure

C_p = external pressure coefficient

(GC_{pi}) = internal pressure coefficient

Gust Effect Factor, G

Dimensions of this building where h/least width = 1.6 < 4.0 indicates that it is a rigid structure,

$$G = 0.925 \left\{ \frac{\left(1 + 1.7\, g_Q I_{\bar{z}} Q\right)}{\left(1 + 1.7\, g_v I_{\bar{z}}\right)} \right\}$$ (Equation 6 - 4)

$$g_Q = g_v = 3.4$$ (6.5.8.1)

$$\bar{z} = 0.6\,(157) = 94.2 \text{ ft (controls)}$$ (6.5.8.1)

$$\bar{z} = z_{min} = 30 \text{ ft}$$ (Table 6 - 4)

$$c = 0.30$$ (Table 6 - 4)

$$I_{\bar{z}} = c\left(\frac{33}{\bar{z}}\right)^{\frac{1}{6}} = 0.30\left(\frac{33}{94.2}\right)^{\frac{1}{6}} = 0.25$$

$$L_{\bar{z}} = \lambda\left(\frac{\bar{z}}{33}\right)^{\bar{\epsilon}} = 320\left(\frac{94.2}{33}\right)^{\frac{1}{3}} = 454 \text{ ft}$$ (Table 6 - 4)

$$Q = \sqrt{\frac{1}{1 + 0.63\left(\dfrac{B+h}{L_{\bar{z}}}\right)^{0.63}}}$$ (Equation 6 - 4)

$B = 100$ ft (smaller value gives larger G)

$$Q = \sqrt{\frac{1}{1 + 0.63\left(\dfrac{100+157}{454}\right)^{0.63}}} = 0.83$$

$$G = 0.925 \left\{ \frac{\left(1 + 1.7 \times 3.4 \times 0.25 \times 0.83\right)}{\left(1 + 1.7 \times 3.4 \times 0.25\right)} \right\} = 0.83$$

Wall External Pressure Coefficients, C_p

The values for the external pressure coefficients for the various wall surfaces are obtained from Figure 6-3.

The windward wall pressure coefficient is 0.8.

The side wall pressure coefficient is -0.7.

The leeward wall pressure coefficient is a function of the L/B ratio. For wind normal to the 200 ft face, L/B = 100/200 = 0.5; therefore, the leeward wall pressure coefficient is -0.5. For wind normal to 100 ft face, L/B = 200/100 = 2.0; therefore, the leeward wall pressure coefficient is -0.3.

Table 3.3.2 Wall C_p

Surface	Wind Direction	L/B	C_p
Windward Wall	All	All	0.80
Leeward Wall	⊥ to 200 ft face	0.5	-0.50
	‖ to 200 ft face	2.0	-0.30
Side Wall	All	All	-0.70

Roof C_p (with the wind normal to the 200 ft face)

For h/L = 157/100 ≈ 1.6 > 1.0, and θ < 10 degrees, two zones are specified:
0 to h/2, C_p = -1.3
> h/2, C_p = -0.7

The C_p = -1.3 may be reduced with the area over which it is applicable.
Area = 200 x 79 = 15,800 sq ft
Reduction factor = 0.8
Reduced C_p = 0.8 x – 1.3 = -1.04

Table 3.3.3 Roof C_p for Wind Normal to 200 ft Face

Distance from Leading Edge	C_p
0 to h/2	-1.04
>h/2	-0.70
* h = 157 ft	

Roof C_p (with the wind normal to 100 ft face)

For h/L = 157/200 ≈ 0.8, interpolation in Figure 6-3 is required.

Table 3.3.4 Roof C_p for Wind Normal to 100 ft Face

Distance from Windward Edge	h/L ≤ 0.5	h/L = 0.8	h/L ≥ 1.0
0 to h/2	-0.9	-0.98	-1.04
h/2 to h	-0.9	-0.78	-0.7
h to 2h	-0.5	-0.62	-0.7

Internal Pressure Coefficients, (GC_{pi})

The building is in a hurricane-prone region. Since the glazing is assumed to be debris resistant in the bottom 60 ft, glazing is not considered as opening. The building is not classified as partially open building.

For enclosed buildings:
$$GC_{pi} = \pm\, 0.18 \qquad \text{(Table 6-7)}$$

MWFRS Pressures
$$p = qGC_p - q_h(GC_{pi}) \qquad \text{(Table 6-1)}$$

Windward Wall Calculation at 30 ft, Wind Normal to 200 ft Face
p = 21.9(0.83)(0.80) − 35.1($\pm$ 0.18)
p = 8.2 psf with (+) internal pressure
p = 20.9 psf with (-) internal pressure

Roof Calculation for 0-79 ft (h/2) from Edge, Wind Normal to 200 ft Face
p =   35.1(0.83)(-1.04) − 35.1($\pm$ 0.18)
  = - 36.6 psf with (+) internal pressure
  = - 24.0 psf with (-) internal pressure

Table 3.3.5 External Pressures for MWFRS: Wind Normal to 200 ft Face

Surface	z	q	C_p	Ext. Press.
	ft	psf		
Windward	0-15	17.9	0.80	11.9
Wall	30	21.9	0.80	14.5
	50	25.4	0.80	16.9
	80	29.1	0.80	19.3
	120	32.6	0.80	21.7
	157	35.1	0.80	23.3
Leeward	All	35.1	-0.50	-14.6
Wall				
Side Walls	All	35.1	-0.70	-20.4
Roof	0-79	35.1	-1.04	-30.3
	79-100	35.1	-0.70	-20.4

Notes: $q_h = 35.1$ psf; G = 0.83

Parapet load on MWFRS

For parapet load contribution to the MWFRS, the parapet may be approximated as a sign structure at ground level.

Force coefficient for $v = 3/200 < 3$, $C_f = 1.2$ (Table 6-11)

For open building and other structures, use the following equation:

$$F = q_z \, GC_f \, A_f \qquad\qquad (\text{Equation 6-20})$$

Note: for signs, $K_d = 0.85$; therefore, $q_z = 35.4$

If A_f is taken as 1 ft wide x 3 ft high,

$$F = 35.4 \, (0.83) \, (1.2) \, (3) = 106 \text{ plf}$$

This force is to be applied on windward parapet and to the leeward parapet.

For design of parapet, see the loads on components and cladding.

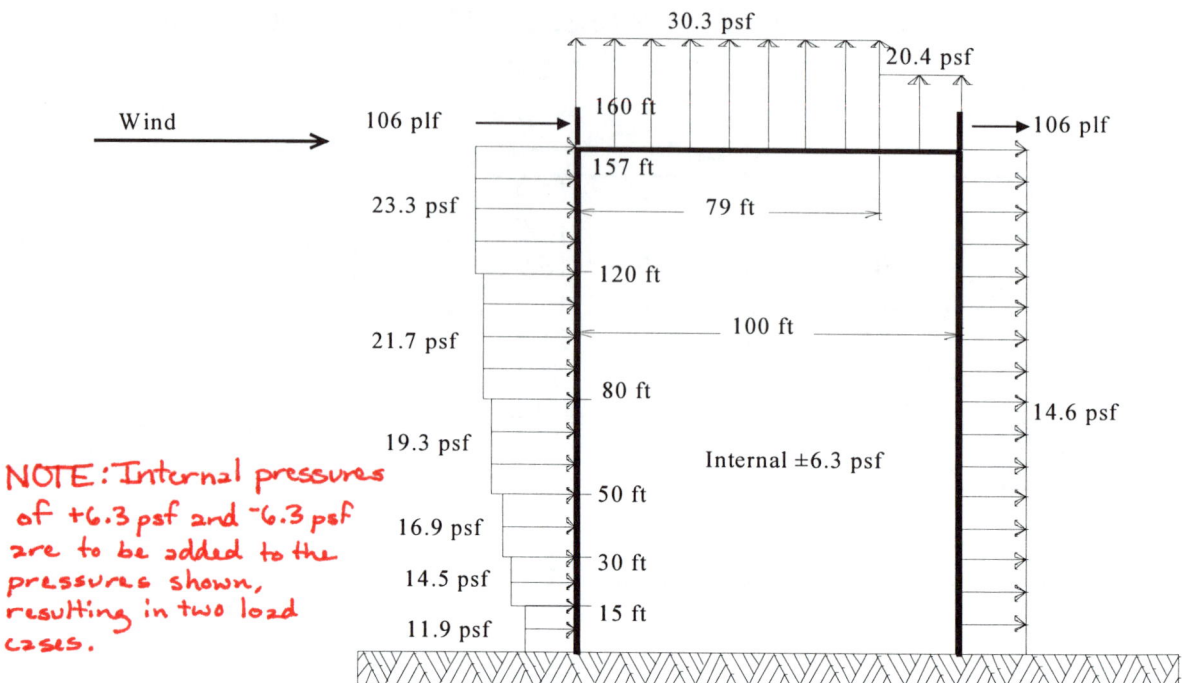

Figure 3.3.2 ~~Net~~ **Pressures for MWFRS for Wind Normal to the 200 ft Face with Positive Internal Pressure**

Table 3.3.6 External Pressures for MWFRS: Wind Normal to 100 ft Face

Surface	z	q	C_p	Ext. Press.
	ft	psf		
Windward	0-15	17.9	0.80	11.9
Wall	30	21.9	0.80	14.5
	50	25.4	0.80	16.9
	80	29.1	0.80	19.3
	120	32.6	0.80	21.7
	157	35.1	0.80	23.3
Leeward Wall	All	35.1	-0.30	-8.7
Side Walls	All	35.1	-0.70	-20.4
Roof	0-79	35.1	-0.98	-28.6
	79-157	35.1	-0.78	-22.7
	157-200	35.1	-0.62	-18.1

Notes: q_h = 35.1 psf; G = 0.83

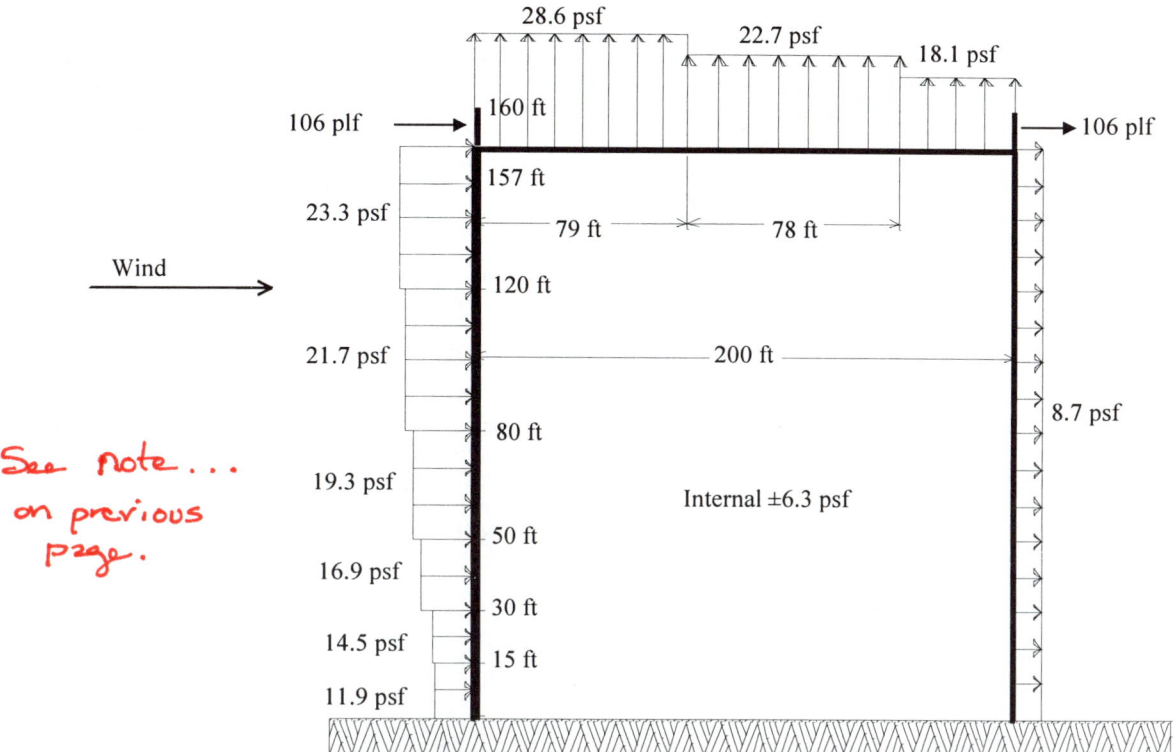

Figure 3.3.3 Pressures for MWFRS for Wind Normal to 100 ft Face

Full and Partial Loading on MWFRS

Section 6.5.12.3 requires that any building with mean roof height h greater than 60 ft be subjected to the four load cases shown in Figure 6-9 to account for torsion. Case 1 includes the loadings determined in this example and shown in Figures 3.3.2 and 3.3.3. A combination of windward (P_W) and leeward (P_L) loads should be applied for load Cases 2, 3, and 4 as shown in Figure 6-9.

Design Pressures for Components and Cladding

Design pressure for components and cladding is obtained by the following equation:

$$p = q(GC_p) - q_i(GC_{pi}) \qquad \text{(Equation 6-19)}$$

where:

$q = q_z$ for windward wall calculated at height z and q_h for leeward wall, side walls, and roof calculated at height h

$q_i = q_h$ for enclosed building

(GC_p) = external pressure coefficient, see Figure 6-8

(GC_{pi}) = internal pressure coefficient, see Table 6-7

Wall Design Pressures

The pressure coefficients (GC_p) are a function of effective wind area. The definition of effective wind area for a component or cladding panel is the span length multiplied by an effective width that need not be less than one-third the span length, see Section 6.2. The effective wind areas, A, for wall components are:

Mullion: larger of
$$A = 11(5) = 55 \text{ sq ft} \quad \text{(controls)}$$
          or
$$A = 11(11/3) = 40.3 \text{ sq ft}$$

Glazing Panel: larger of
$$A = 5(5.5) = 27.5 \text{ sq ft} \quad \text{(controls)}$$
          or
$$A = 5(5/3) = 8.3 \text{ sq ft}$$

Width of Corner Zone 5: larger of
$$a = 0.1(100) = 10 \text{ ft} \quad \text{(controls)}$$
          or
$$a = 3 \text{ ft}$$

Table 3.3.7 Wall (GC_p)

Component	A (sq ft)	Zones 4 and 5 $(+GC_p)$	Zone 4 $(-GC_p)$	Zone 5 $(-GC_p)$
		GC_p		
Mullion	55	0.81	-0.84	-1.55
Panel	27.5	0.87	-0.88	-1.72

The internal pressure coefficient $(GC_{pi}) = \pm 0.18$.   (Table 6-7)

Typical Design Pressure Calculations

Mullion in Zone 4 for side and leeward walls
$$p = 35.1(-0.84) - 35.1(\pm 0.18)$$
$$= -35.8 \text{ psf (positive internal pressure controls)}$$

Controlling negative pressure is obtained with positive internal pressure and controlling positive pressure is obtained with negative internal pressure.

Table 3.3.8 Mullion Design Pressures, psf

| Component | z | Design Pressures | | | |
| | | Zone 4 | | Zone 5 | |
	ft	Positive	Negative	Positive	Negative
Mullion	0-15	24.1	-35.8	24.1	-60.7
	15-30	24.1	-35.8	24.1	-60.7
	30-50	26.9	-35.8	26.9	-60.7
	50-80	29.9	-35.8	29.9	-60.7
	80-120	32.7	-35.8	32.7	-60.7
	120-157	34.8	-35.8	34.8	-60.7

Table 3.3.9 Panel Design Pressures, psf

| Component | z | Design Pressures | | | |
| | | Zone 4 | | Zone 5 | |
	ft	Positive	Negative	Positive	Negative
Panel	0-15	25.4	-37.2	25.4	-66.7
	15-30	25.4	-37.2	25.4	-66.7
	30-50	28.4	-37.2	28.4	-66.7
	50-80	31.6	-37.2	31.6	-66.7
	80-120	34.7	-37.2	34.7	-66.7
	120-157	36.9	-37.2	36.9	-66.7

Parapet Design Pressures

Although the Standard does not provide specific design pressures for parapets, the values of GC_p given in Figure 6-8 allow for rational assessment of these pressures. For a parapet on the windward building face, the net pressure is assumed to be algebraic sum of the positive pressure for walls, Zones 4 and 5, and the negative pressure for roof, Zone 2. For a parapet on the leeward or corner of the building face, the net pressure is assumed to be the algebraic sum of the positive pressure for walls, Zones 4 and 5, and the negative pressure for walls, Zones 4 and 5. The effective wind area will depend upon the dimensions and framing details of the parapet. In this example, the effective wind area is assumed to be 3 ft x 3 ft = 9 sq ft.

Windward Parapet:

$$p = 35.4(0.9) - 35.4(-2.3)$$
$$= 113.3 \text{ psf (Directed inward)}$$

Leeward or Corner Area Parapet:

Zone 4: $p = 35.4(0.9) - 35.4(-0.9)$
                           $= 63.7$ psf (Directed outward)

Zone 5: $p = 35.4(0.9) - 35.1(-1.8)$
                           $= 95.0$ psf (Directed outward)

Roof Design Pressures

The component and cladding roof pressure coefficients are given in Figure 6-8. The pressure coefficients are a function of the effective wind area. Since specific components of roof are not identified, design pressures are given for various effective wind areas, A.

Table 3.3.10 Roof (GC$_p$)

| A | External Pressure Coefficient | |
| | Zone 1 | Zones 2 and 3 |
Sq ft	(-GC$_p$)	(-GC$_p$)*
≤ 10	-1.40	-2.30
20	-1.31	-2.18
100	-1.11	-1.89
250	-0.99	-1.72
400	-0.93	-1.64
≥ 500	-0.90	-1.60

* Note 7 in Figure 6-8 permits treatment of Zone 3
as Zone 2 if parapet of 3 ft or higher is provided.

The design pressures are the algebraic sum of external and internal pressures. Positive internal pressure provides controlling negative pressures. These design pressures act across the roof surface (interior to exterior).

Table 3.3.11 Roof Design Pressures, psf

| A | Design Pressures | |
| | Negative | |
sq ft	Zone 1	Zones 2 and 3
≤ 10	-55.1	-87.1
20	-52.3	-82.8
100	-45.3	-72.7
250	-41.1	-66.7
400	-39.0	-63.9
500	-37.9	-62.5

3.4   Example 4 – Office Building of Example 3 Located on Escarpment

In this example, velocity pressures for the office building of Example 3, when it is located on an escarpment, are determined. Design pressures for MWFRS and components and cladding can be determined in the same manner as Example 3 once velocity pressures q_z and q_h are determined.

Location:	City in Alaska
Topography:	Escarpment as shown
Terrain:	Suburban
Dimensions:	100 ft x 200 ft in plan
	Roof height of 157 ft with 3 ft parapet
	Flat roof
Framing:	Reinforced concrete rigid frame in both directions
	Floor and roof slabs provide diaphragm action
	Fundamental natural frequency is greater than 1 Hz
Cladding:	Mullions for glazing panels span 11 ft between floor slabs
	Mullion spacing is 5 ft
	Glazing panels are 5 ft wide x 5 ft 6 in. high (typical). Glazing does not have to be wind-borne debris impact resistant because Alaska is not in a hurricane-prone region, see Section 6.5.9.3.

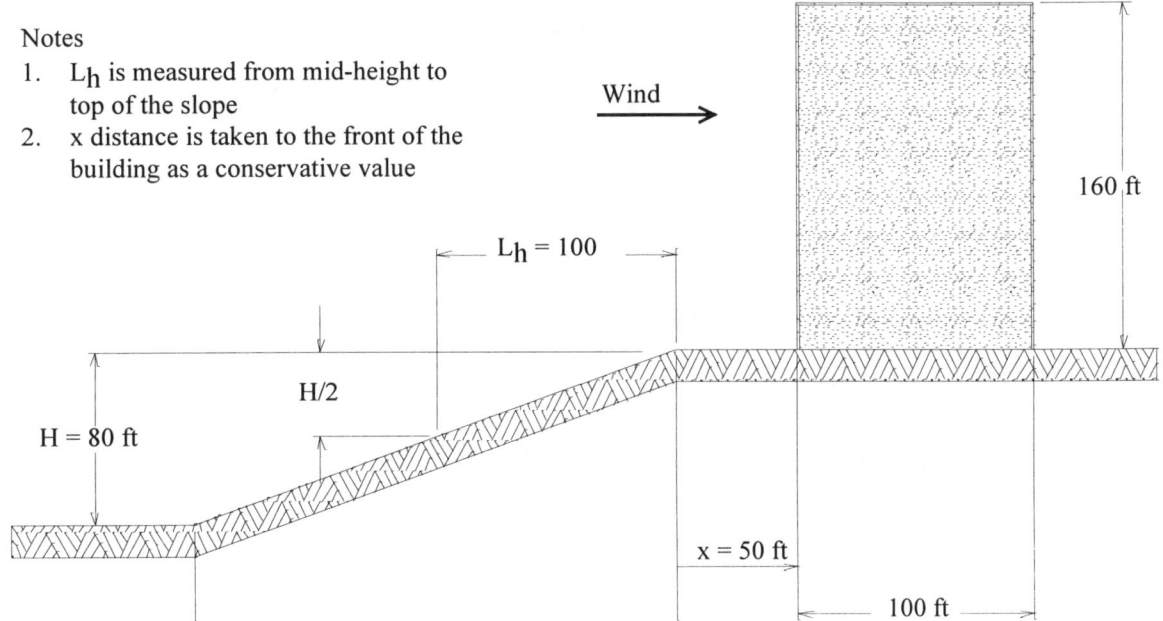

Notes
1. L_h is measured from mid-height to top of the slope
2. x distance is taken to the front of the building as a conservative value

Wind

$L_h = 100$

160 ft

H/2

H = 80 ft

x = 50 ft

100 ft

Figure 3.4.1 Office Building on Escarpment

Exposure, Building Classification, and Basic Wind Speed

Same as Example 3: Exposure B
Category II
$V = 120$ mph, same as Example 3

Velocity Pressures

The velocity pressure equation is:

$q_z = 0.00256 K_z K_{zt} V^2 I$ psf

For this example, K_z is obtained from Table 6-5 and K_{zt} is determined using Figure 6-2. $I = 1.0$ for Category II and $V = 120$ mph.

Determination of K_{zt}

The topographic effect of escarpment applies only when the upwind terrain is free of topographic features for a distance equal to 100H or 2 miles, whichever is smaller. For this example, it is assumed that there are no topographic features upwind for a distance of 8000 ft.

For use in Figure 6-2 of the Standard:
$H = 80$ ft
$L_h = 100$ ft
$x = 50$ ft (distance to the front face of the building)

Since $\dfrac{H}{L_h} = 0.8 > 0.5$, according to Note 2 in Figure 6-2

use $\dfrac{H}{L_h} = 0.5$ and $L_h = 2H = 160$ ft

The building is on a 2-D escarpment.

For Exposure B, $K_1 = (0.75)(0.5) = 0.38$ (Figure 6-2)

For $\dfrac{x}{L_h} = 0.31$; $K_2 = \left(1 - \dfrac{0.31}{4}\right) = 0.92$ (Figure 6-2)

$K_3 = e^{-2.5z/L_h}$ (values in table for z)

$K_{zt} = (1 + K_1 K_2 K_3)^2$ (Equation 6-2)

$q_z = 0.00256 K_z K_{zt}(120)^2(1.0)$

Table 3.4.1 Speed-up Velocity Pressures, psf

Height, ft	K_z	z/L_h*	K_3	K_{zt}	q_z, psf
0-15	0.57	0.05	0.88	1.71	30.5
30	0.70	0.14	0.71	1.56	34.2
50	0.81	0.25	0.54	1.41	35.8
80	0.93	0.41	0.36	1.27	37.0
120	1.04	0.63	0.21	1.15	37.5
h = 157	K_h = 1.12	0.87	0.11	1.08	37.9

* z is taken midway between the height range because it is unconservative for K_{zt} to take top height of the range.
Note; L_h = 160 ft

Effect of Escarpment

Velocity pressures q_z are compared with the values of Example 3 to assess the effect of the escarpment. The increase in velocity pressures does not directly translate into an increase in design pressures as discussed below.

Table 3.4.2 Velocity Pressure q_z, psf

Height, ft	Homogeneous Terrain Example 3	Escarpment Example 4	% Increase
0-15	17.9	30.5	71
30	21.9	34.2	56
50	25.4	35.8	41
80	29.2	37.0	27
120	32.6	37.5	15
157 (roof)	35.1	37.9	8

For MWFRS, the windward wall pressures will increase by the percentages shown at various heights; however the leeward wall, side wall, roof, and internal pressures will increase by 10% since these pressures are controlled by velocity pressure at roof height, q_h.

For components and cladding, the negative (outward acting) design pressures will also increase only by 10%.

3.5   Example 5 – 2500 sq ft House with Gable/Hip Roof

Design wind pressures for a typical one-story house are to be determined. The physical data are as follows:

Location:	Dallas–Fort Worth, Texas
Topography:	Homogenous
Terrain:	Suburban
Dimensions:	80 ft x 40 ft (including porch) footprint
	Porch is 8 ft x 48 ft
	Wall eave height is 10 ft
	Roof gable Θ = 15 degrees; roof overhang is 2 ft all around
Framing:	Typical timber construction
	Wall studs are spaced 16 in. on center
	Roof trusses spanning 32 ft are spaced 4 ft on center
	Roof panels are 4 ft x 8 ft
	Glazing is uniformly distributed (pressures on C&C will depend on effective area and location; all items are not included for brevity)

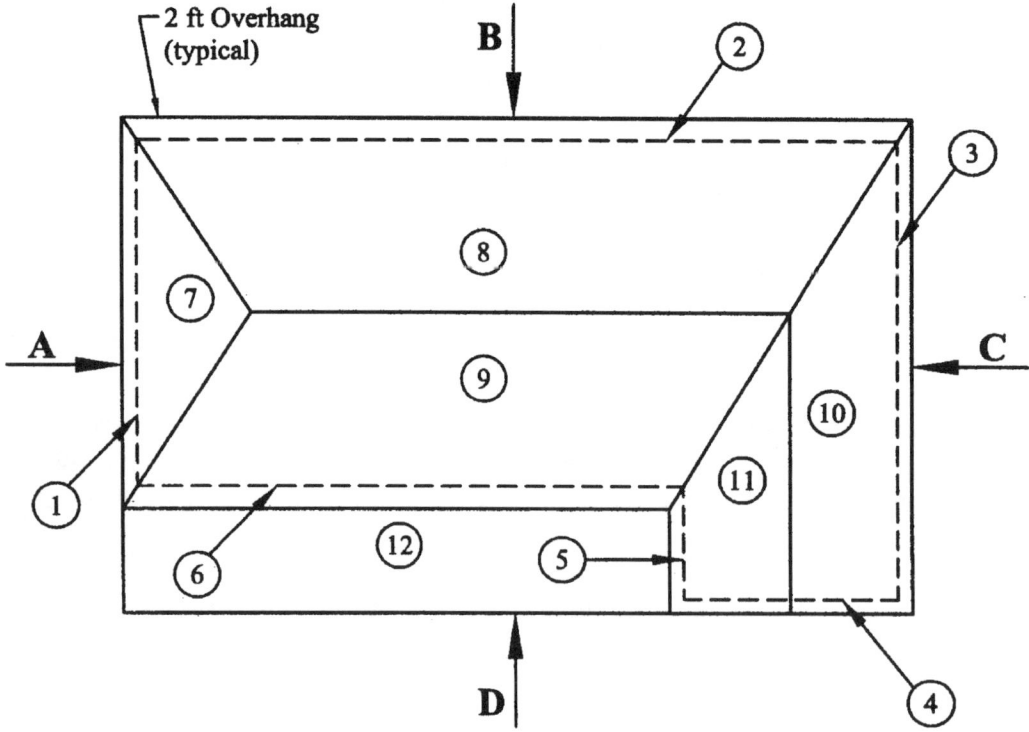

Figure 3.5.1 View of roof of 2500 sq ft house

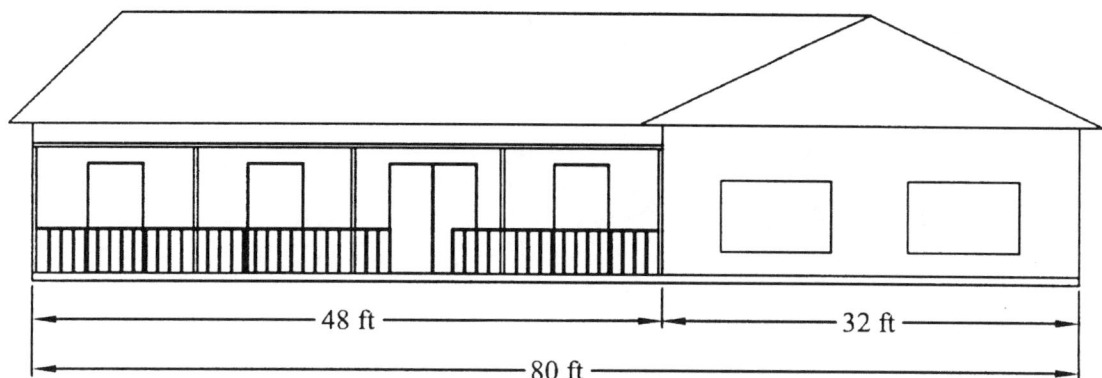

Figure 3.5.2 Front view

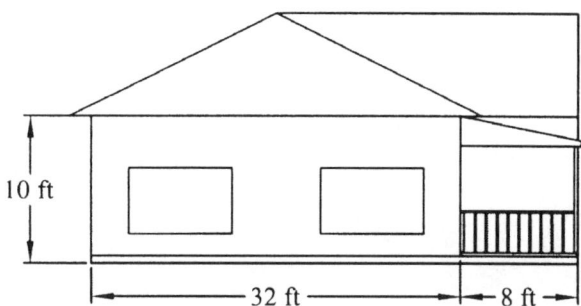

Figure 3.5.3 Side view

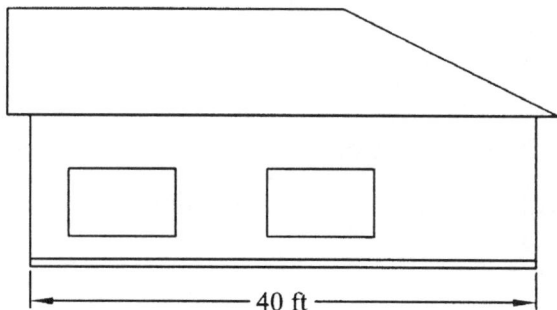

Figure 3.5.4 Side view

Wind speed V = 90 mph

Importance factor I = 1.0

Topography factor K_{zt} = 1.0

Directionality factor K_d = 0.85 (for buildings)

Terrain exposure coefficient depends on terrain, see Exposure B, and on h

$$\text{mean roof height} = 10 + \frac{(16)(\tan 15°)}{2} = 12.1 \text{ ft}$$

Since K_z is constant in 0-15 ft region, from Table 6-5,
$K_z = K_h = 0.70$ for Case 1 (C&C)
· $K_z = K_h = 0.57$ for Case 2 (MWFRS)

Velocity Pressures

$$q_z = 0.00256 \, K_z \, K_{zt} \, G \, V^2 \, I \text{ psf} \qquad \text{(Equation 6-1)}$$

For MWFRS, $q_z = q_h = 0.00256 \, (0.57)\,(1.0)\,(0.85)\,(90)^2\,(1.0)$
$= 10.1$ psf

For C&C, $q_z = q_h = 0.00256 \, (0.7)\,(1.0)\,(0.85)\,(90)^2\,(1.0)$
$= 12.3$ psf

Gust Effect Factor

$G = 0.85$ \qquad (Section 6.5.8.1)

$(GC_{pi}) = +0.18$ and -0.18 \qquad (Table 6-7)

Wind Pressure for MWFRS

Because of asymmetry, all four wind directions are considered (normal to walls)

The wall surfaces are numbered 1 through 6; roof surfaces are 7 through 11; porch roof surface is 12.

WIND DIRECTION A

Wall pressures:

Surface 1: $p = 10.1\,(0.85)\,(0.8) - 10.1\,(\pm 0.18) = +6.9 \pm 1.8$ psf (windward)

Surface 2: $p = 10.1\,(0.85)\,(-0.7) - 10.1\,(\pm 0.18) = -6.0 \pm 1.8$ psf (side)

Surface 3: $p = 10.1\,(0.85)\,(-0.3) - 10.1\,(\pm 0.18) = -2.6 \pm 1.8$ psf (leeward)

(for $L/B = 80/40 = 2$; $C_p = -0.3$)

Surface 4: $p = -6.0 \pm 1.8$ psf (side)

Surface 5: $p = +6.9 \pm 1.8$ psf (windward)

Surface 6: $p = -6.0 \pm 1.8$ psf (side)

Roof pressures: $h/L = 12.1/80 = 0.15$; $\Theta = 15$ degrees

Surface 7: $p = 10.1\,(0.85)\,(-0.5) - 10.1\,(\pm 0.18) = -4.3 \pm 1.8$ psf (windward)

Surface 8: for $\Theta = 0$ degrees; pressure varies along the roof

$p = 10.1\,(0.85)\,(-0.9) - 10.1\,(\pm 0.18) = -7.7 \pm 1.8$ psf; 1 to 12.1 ft

$p = 10.1\,(0.85)\,(-0.5) - 10.1\,(\pm 0.18) = -4.3 \pm 1.8$ psf; 12.1 to 24.2 ft

$p = 10.1\,(0.85)\,(-0.3) - 10.1\,(\pm 0.18) = -2.6 \pm 1.8$ psf; 24.2 ft to end

Surface 9: same pressures as surface 8

Surface 10: $p = 10.1\,(0.85)\,(-0.5) - 10.1\,(\pm 0.18) = -4.3 \pm 1.8$ psf (leeward)

Surface 11: $p = 10.1\,(0.85)\,(-0.5) - 10.1\,(\pm 0.18) = -4.3 \pm 1.8$ psf (windward)

Surface 12: same as surface 8 without internal pressure

Overhang pressures: At wall surfaces 1 and 5

$p = 10.1\,(0.85)\,(0.8) = +6.9$ psf

Internal pressure is of the same sign on all applicable surfaces

WIND DIRECTION B

<u>Wall pressures:</u>

Surface 1: $p = -6.0 \pm 1.8$ psf (side)

Surface 2: $p = +6.9 \pm 1.8$ psf (windward)

Surface 3: $p = -6.0 \pm 1.8$ psf (side)

Surface 4: $p = 10.1 \, (0.85) \, (-0.5) - 10.1 \, (\pm 0.18) = -4.3 \pm 1.8$ psf (leeward)

(for $L/B = 40/80 = 0.5$; $C_p = -0.5$)

Surface 5: even though technically this surface is side wall, it is likely to see the same pressure as surface 6

Surface 6: same pressure as surface 4

<u>Roof pressures:</u> $h/L = 12.1/40 = 0.3$; $\Theta = 15$ degrees

For windward, $C_p = -0.54$ (interpolated)

For leeward, $C_p = -0.5$

For parallel to ridge, $C_p = -0.9, -0.5,$ and -0.3

Surface 7: same pressures as surface 8 for Wind Direction A

Surface 8: $p = 10.1 \, (0.85) \, (-0.54) - 10.1 \, (\pm 0.18) = -4.6 \pm 1.8$ psf (windward)

Surface 9: $p = 10.1 \, (0.85) \, (-0.5) - 10.1 \, (\pm 0.18) = -4.3 \pm 1.8$ psf (leeward)

Surface 10: same pressures as surface 8 for Wind Direction A

Surface 11: same as surface 9 because it is sloping with respect to ridge

Surface 12: This surface is at a distance greater than 2h

$p = 10.1 \, (0.85) \, (-0.3) = -2.6$ psf; no internal pressure

<u>Overhang pressures:</u> At wall surface 2

$p = 10.1 \, (0.85) \, (0.8) = +6.9$ psf

Internal pressure is of the same sign on all applicable surfaces

WIND DIRECTION C

<u>Wall pressures:</u>

Surface 1 and 5: $p = -2.6 \pm 1.8$ psf (leeward)

Surface 2, 4, and 6: $p = -6.0 \pm 1.8$ psf (side)

Surface 3: $p = +6.9 \pm 1.8$ psf (windward)

<u>Roof pressures:</u>

Surface 7 and 11: $p = -4.3 \pm 1.8$ psf (leeward)

Surface 8 and 9: pressures vary along the roof; same pressures as surface 8 for Wind

Direction A

Surface 10: $p = -4.3 \pm 1.8$ psf (windward)

Surface 12: same pressures as surface 9 without internal pressures

<u>Overhang pressures:</u> At wall surface 3

$$p = 10.1 \, (0.85) \, (0.8) = +6.9 \text{ psf}$$

Internal pressure is of the same sign on all applicable surfaces.

WIND DIRECTION D

<u>Wall pressures:</u>

Surface 1 and 3: p = -6.0 $\pm$ 1.8 psf (side)

Surface 2: p = -4.3 $\pm$ 1.8 psf (leeward)

Surface 4, 5, and 6: p = +6.9 $\pm$ 1.8 psf (windward)

<u>Roof pressures:</u>

Surface 7, 10, and 11: pressures vary along the roof; same pressures as surface 8 for Wind Direction A

Surface 8: p = -4.3 $\pm$ 1.8 psf (leeward)

Surface 9: p = -4.3 $\pm$ 1.8 psf (windward)

Surface 12: This surface will see pressures on top and bottom surfaces; they will add algebraically.

For $\Theta = 0°$, h/L < 0.5, $C_p = -0.9$

p = 10.1 (0.85) (-0.9) – 10.1 (0.85) (+0.8) = -14.6 psf uplift

<u>Overhang pressures:</u> At wall surfaces 4, 5, and 6

p = 10.1 (0.85) (0.8) = +6.9 psf

Internal pressure is of the same sign on all applicable surfaces.

Components and Cladding

Wall Component:

    Wall studs are 10 ft long and spaced 16 in. apart.

    Effective area = larger of 10 x 1.33 = 13.3 sq ft

                or 10 x 10/3 = 33.3 sq ft (controls)

    From Figure 6-5a, Equations in Chapter 2 are used

    (GC_p) = +0.91 for Zones 4 and 5

    (GC_p) = -1.01 for Zone 4

    (GC_p) = -1.22 for Zone 5

    Distance 'a' = smaller of 0.1 (40) = 4 ft (controls)

                or 0.4 (12.1) = 4.8 ft

    Design pressure p = 12.3 (0.91 + 0.18) = +13.4 psf (all walls)

        p = 12.3 (-1.01 – 0.18) = -14.6 psf (middle)

        p = 12.3 (-1.22 – 0.18) = -17.2 psf (corner)

Roof component:

    Roof trusses are 32 ft long and spaced 4 ft apart.

    Effective area = larger of 32 x 4 = 128 sq ft

                or 32 x 32/3 = 341 sq ft (controls)

    From Figure 6-5b for $\Theta = 15°$

    (GC_p) = +0.3 for Zones 1, 2, and 3

    (GC_p) = -0.8 for Zone 1

    (GC_p) = -1.4 for Zones 2 and 3

Distance 'a' = smaller of 0.1 (40) = 4 ft (controls)

or 0.4 (12.1) = 4.8 ft

Design pressures

$p = 12.3 (0.3 + 0.18) = +5.9$ psf (all zones)

$p = 12.3 (-0.8 - 0.18) = -12.1$ psf (middle roof)

$p = 12.3 (-1.4 - 0.18) = -19.4$ psf (edges of roof)

Overhang pressures to be used for reaction and anchorage

$p = 12.3 (-2.2 - 0.18) = -29.3$ psf (edge of roof)

$p = 12.3 (-2.5 - 0.18) = -33.0$ psf (roof corners)

Roof Panels

Effective area = 4 x 8 = 32 sq ft

From Figure 6-5b for $\Theta = 15°$ (note: Zones 2 and 3 are overhang)

$(GC_p) = +0.4$ for Zones 1, 2, and 3

$(GC_p) = -0.85$ for Zone 1

$(GC_p) = -2.2$ for Zones 2 (with overhang)

$(GC_p) = -3.1$ for Zone 3 (with overhang)

Distance 'a' = smaller of 0.1 (40) = 4 ft (controls)

or 0.4 (12.1) = 4.8 ft

Design pressures

$p = 12.3 (0.4 + 0.18) = +7.1$ psf (all zones)

$p = 12.3 (-0.85 - 0.18) = -12.7$ psf (middle roof)

$p = 12.3 (-2.2 - 0.18) = -29.3$ psf (edges of roof)

$p = 12.3 (-3.1 - 0.18) = -40.3$ psf (roof corners)

3.6   Example 6 – House of Example 5 on an Isolated Hill

In this example, the one-story house of Example 5 is placed on a hill to illustrate the effect of topography. It is only necessary to calculate velocity pressure, q_z, and compare with the value of Example 5.

The data for the house are the same as Example 5. The parameter of significance is the mean roof height, which is $h = 12.1$ ft.

The topography dimensions of a 3-D axisymmetrical hill are as shown:

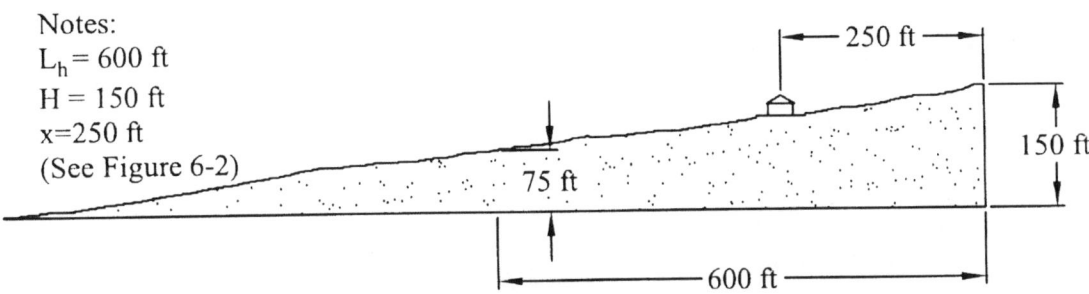

Notes:
$L_h = 600$ ft
$H = 150$ ft
$x = 250$ ft
(See Figure 6-2)

Figure 3.6.1 One-Story House on a Hill

Basic Wind Speed, Exposure, Classification

Same as Example 5:   Basic wind speed, $V = 90$ mph
Category II, $I = 1.0$
Exposure B, $K_z = K_h = 0.57$ for Case 2 (MWFRS)
$K_h = 0.7$ for Case 1 (C&C)
Directionality factor, $K_d = 0.85$
Classification: Enclosed

Determination of K_{zt}

The topographic effect is applicable only when the following conditions are met, see Section 6.5.7.1:

1. The hill is isolated and unobstructed upwind by similar topographic features for a distance of the larger of 100 x H = 15,000 ft (controls) or two miles.
2. There are no hills higher than 75 ft for a distance of two miles.
3. The building is located in the upper one-half of the hill.
4. $H/L_h = 150/600 = 0.4 \geq 0.2$
5. $H = 150$ ft > 60 ft for Exposure B

$\qquad$ Topographic factor $K_{zt} = (1 + K_1 K_2 K_3)^2$ $\qquad\qquad$ (Equation 6-1)

$\qquad$ For $H/L_h = 0.4$, x = 250 ft and Exposure B $\qquad\qquad$ (Figure 6-2)

$\qquad$ $K_1 = (0.95)(0.4) = 0.38$

$$K_2 = \left(1 - \frac{250}{(1.5)(600)}\right) = 0.72$$

$$K_3 = e^{-4(15)/600} = 0.90 \qquad (z \text{ is taken as 15 ft to be consistent with } K_z)$$

$\qquad$ $K_{zt} = [1 + (0.38)(0.72)(0.90)]^2 = 1.56$

Velocity Pressure

$\qquad\qquad$ $q_z = 0.00256 K_z K_{zt} K_d V^2 I$ $\qquad\qquad$ (Equation 6-13)

$\qquad\qquad$ $= 0.00256\,(0.57)\,(1.56)\,(0.85)\,(90)^2\,(1.0)$

$\qquad\qquad$ $= 15.7$ psf

Assessment of Topographic Effect

The velocity pressure (hence, design pressure) increases by 56% (15.7 psf versus 10.1 psf) for MWFRS when the house is located on the hill. The velocity pressure increase for C&C will also be 56%. The increase would be 80% if the house were located at the crest of the hill.

3.7 Example 7 – 200 ft x 250 ft Gable Roof Commercial/Warehouse Building Using Rigid Buildings of All Height Provisions

In this example, design wind pressures for a large, one-story commercial-industrial building are determined. The building data are as follows:

Location: Memphis, Tennessee
Terrain: Flat farmland
Dimensions: 200 ft × 250 ft in plan
Eave height of 20 ft
Roof Slope 4:12 (18.4 degrees)
Framing: Rigid frames span the 200 ft direction
Rigid frame bay spacing is 25 ft
Lateral bracing in the 250 direction is provided by a "wind truss" spanning the 200 ft to side walls and cable/rod bracing in the planes of the walls
Girts and purlins span between rigid frames (25 ft span)
Girt spacing is 6 ft 8 in.
Purlin spacing is 5 ft
Cladding: Roof panel dimensions are 2 ft wide
Roof fastener spacing on purlins is 1 ft on center
Wall panel dimensions are 2 ft × 20 ft
Wall fastener spacing on girts is 1 ft on center
Openings are uniformly distributed

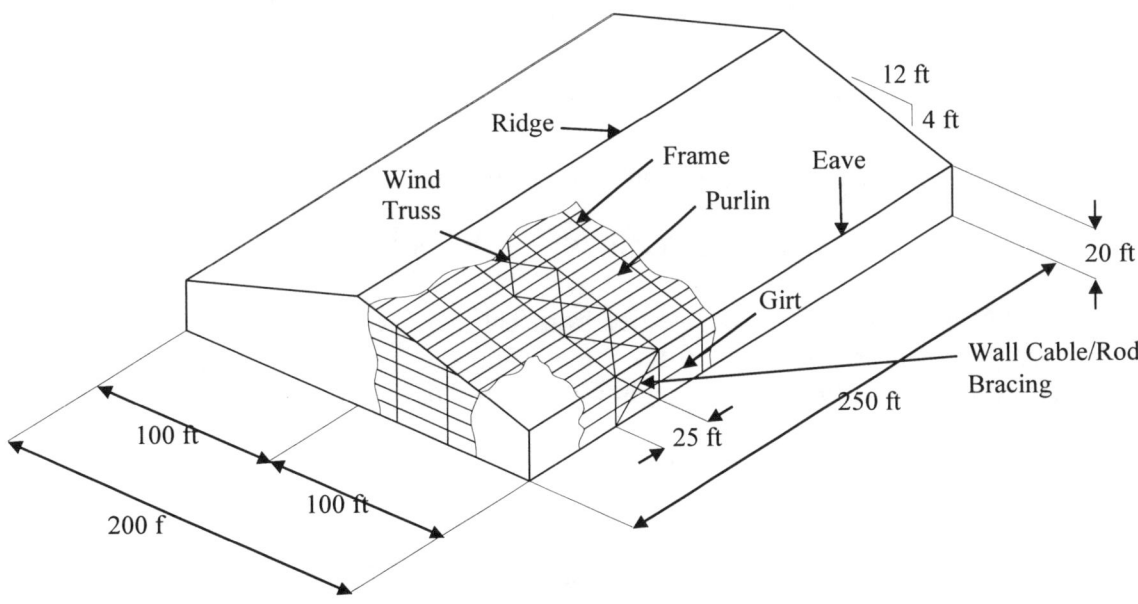

Figure 3.7.1 Dimensions and Framing of the Building of Example 7

Exposure and Building Classification

The building is located in flat, open farmland; therefore, Exposure Category C is applicable, see Section 6.5.6.

The building function is commercial-industrial. It is not considered an essential facility or likely to be occupied by 300 persons at one time. Category II is appropriate, Table 1-1 and Table 6-1 specify an importance factor I = 1.0 (50-year mean return interval).

Basic Wind Speed

Selection of basic wind speed is addressed in Section 6.5.4 of the Standard. Memphis, Tennessee, is neither located in the special wind region nor is there any reason to suggest that winds at the site are unusual and require additional attention. Therefore, the basic wind speed is V = 90 mph, see Figure 6-1.

Analytical Procedure

The building does not qualify for simplified procedure of Section 6.4 because roof slope θ > 10 degrees. *Method 2 – Analytical Procedure will be used,* see Section 6.5.

Wind Directionality

Wind directionality factor is given in Table 6-6. For MWFRS and C&C the factor $K_d = 0.85$.

Velocity Pressures

The velocity pressures are computed using Equation 6-13 of the Standard:

$$q_z = 0.00256 \, K_z K_{zt} K_d V^2 I \quad \text{(psf)}$$

For this example, K_z is obtained from Table 6-5, $K_{zt} = 1.0$ (no topographic effect), I = 1.0 for Category II building, $K_d = 0.85$, and the basic wind speed is V = 90 mph. Substituting these values into the Equation 6-13 yields:

$$q_z = 0.00256 K_z (1.0)(0.85)(90)^2 (1.0)$$
$$q_z = 17.6 \, K_z \quad \text{(psf)}$$

Values for K_z are the same for Cases 1 and 2 for Exposure C, see Table 6-5. Mean roof height h = 36.7 ft.

Table 3.7.1 Velocity Pressures, psf

Height	Ft	K_z	q_z, psf
	0-15	0.85	15.0
Eave ht	20	0.90	15.8
	30	0.98	17.3
h	36.7	1.02	18.0*
	40	1.04	18.3
	50	1.09	19.2
Ridge ht	53.3	1.10	19.4

* $q_h = 18.0$ psf

Design Wind Pressure

Design wind pressures for MWFRS of this building can be obtained using Section 6.5.12.2.1 buildings of all heights or Section 6.5.12.2.2 low-rise building. Pressures determined in this example are using buildings of all heights criteria. Example 8 illustrates use of low-rise building criteria.

$$p = qGC_p - q_i(GC_{pi}) \qquad \text{(Equation 6-15)}$$

where
q = q_z for windward wall at height z above ground
q = q_h for leeward wall, side walls, and roof
q_i = q_h for enclosed buildings
G = gust effect factor
C_p = values obtained from Figure 6-3
(GC_{pi}) = values obtained from Table 6-7

For this example, when the wind is normal to the ridge, the windward roof experiences both positive and negative external pressures. Combining these external pressures with positive and negative internal pressures will result in four loading cases when wind is normal to the ridge.

When wind is parallel to the ridge, positive and negative internal pressures result in two loading cases. The external pressure coefficients C_p for $\theta = 0$ degrees apply in this case.

Gust Effect Factor

For rigid structures, G can be calculated using Equation 6-2, see 6.5.8.1 or alternatively taken as 0.85. G = 0.85 is used in this example for simplicity.

External Wall C_p from Figure 6-3

The pressure coefficients for the windward wall and for the side walls are 0.8 and -0.7, respectively, for all L/B ratios.

The leeward wall pressure coefficient is a function of the L/B ratio. For wind normal to the ridge, L/B = 200/250 = 0.8; therefore, the leeward wall pressure coefficient is -0.5. For flow parallel to the ridge, L/B = 250/200 = 1.25; the value of C_p is obtained by linear interpolation. The wall pressure coefficients are summarized below.

Table 3.7.2 Wall C_p

Surface	Wind Direction	L/B	C_p
Windward Wall	All	All	0.80
Leeward Wall	⊥ to ridge	0.8	-0.50
	‖ to ridge	1.25	-0.45*
Side Wall	All	All	-0.70

* by linear interpolation

Roof C_p from Figure 6-3 (Wind Normal to Ridge)

The roof pressure coefficients for the MWFRS are obtained from the lower table in Figure 6-3. For the roof angle of 18.4 degrees, linear interpolation is used to establish C_p. For wind normal to the ridge, h/L = 36.7/200 = 0.18; hence, only single linear interpolation is required. Note that interpolation is only carried out between values of the same sign.

Table 3.7.3 Roof C_p (Wind Normal to Ridge)

Surface	15°	18.4°	20°
Windward Roof	-0.5	-0.36*	-0.3
	0.0	0.14*	0.2
Leeward Roof	-0.5	-0.57*	-0.6

* by linear interpolation

Internal GC_{pi}

Values for GC_{pi} for buildings are addressed in Section 6.5.11.1 and Table 6-7 of the Standard.

The openings are evenly distributed in the walls (enclosed building) and Memphis, Tennessee is not in a hurricane-prone region. Reduction factor of Section 6.5.11.1.1 (pg 30) is not applicable for enclosed buildings; therefore,

$$GC_{pi} = \pm 0.18$$

MWFRS Net Pressures

$$p = qGC_p - q_h(GC_{pi}) \qquad \text{(Equation 6-15)}$$
$$p = q(0.85)C_p - 18.0(\pm 0.18)$$
where
$q = q_z$ for windward wall
$q = q_h$ for leeward wall, side wall, and roof

Typical Calculation

Windward Wall, 0-15 ft, Wind Normal to Ridge

$p = 15.0(0.85)(0.8) - 18.0(\pm 0.18)$
$p = 7.0$ psf with (+) internal pressure
$p = 13.4$ psf with (-) internal pressure

The net pressures for the MWFRS are summarized in the following tables.

Table 3.7.4 MWFRS Pressures: Wind Normal to Ridge

Surface	z (ft)	q (psf)	G	C_p	Net Pressure psf With	
					(+GC_{pi})	(-GC_{pi})
Windward Wall	0-15	15.0	0.85	0.8	7.0	13.4
	20	15.8	0.85	0.8	7.5	14.0
Leeward Wall	All	18.0	0.85	-0.5	-10.9	-4.4
Side Walls	All	18.0	0.85	-0.7	-14.0	-7.5
Windward	-	18.0	0.85	-0.36	-8.8	-2.3
Roof*				0.14	-1.1	5.4
Leeward Roof	-	18.0	0.85	-0.57	-12.0	-5.5

Notes: $q_h = 18.0$ psf; $(GC_{pi}) = \pm 0.18$; $q_h(GC_{pi}) = \pm 3.2$ psf
    * Two loadings on windward roof and two internal pressures yield a total of four loading cases, see Figures 3.7.2 and 3.7.3.

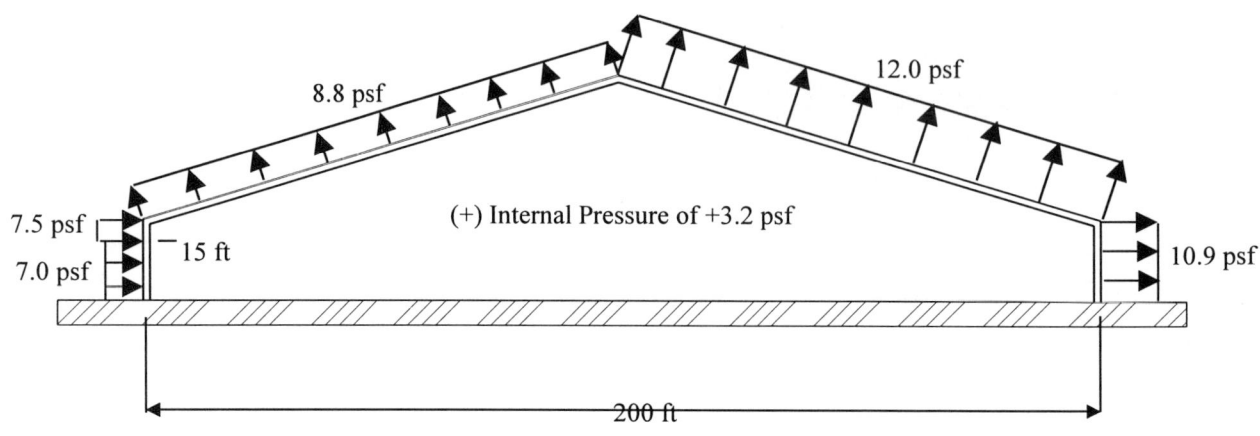

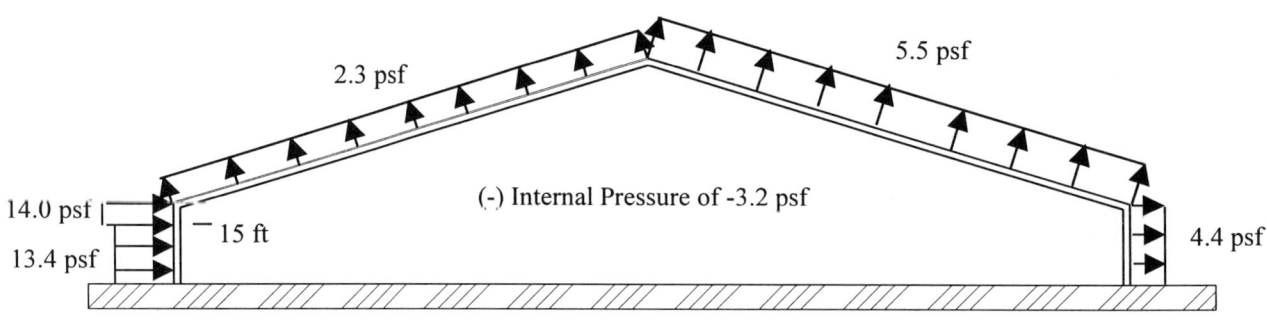

Figure 3.7.2 Net Design Wind Pressures for MWFRS when Wind is Normal to Ridge with Negative Windward External Roof Pressure Coefficient

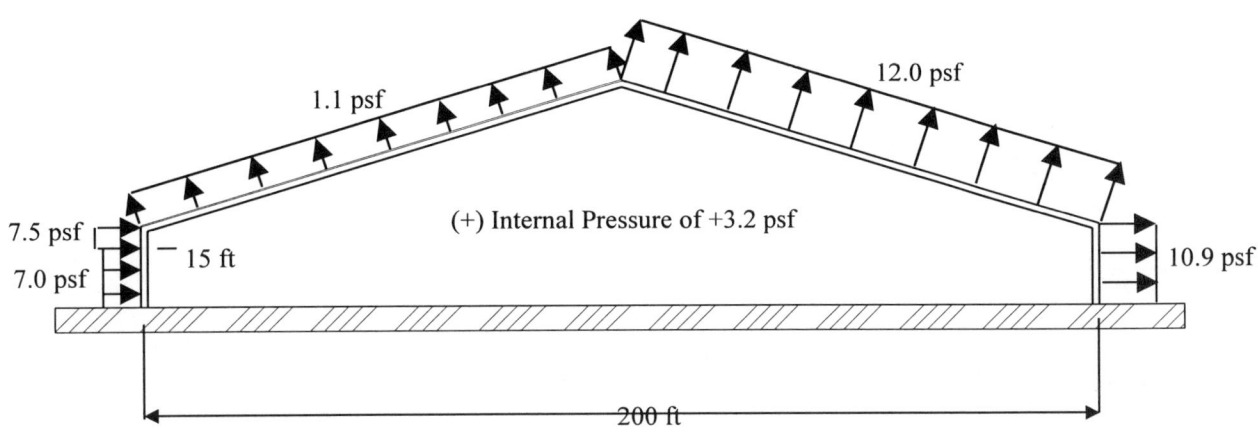

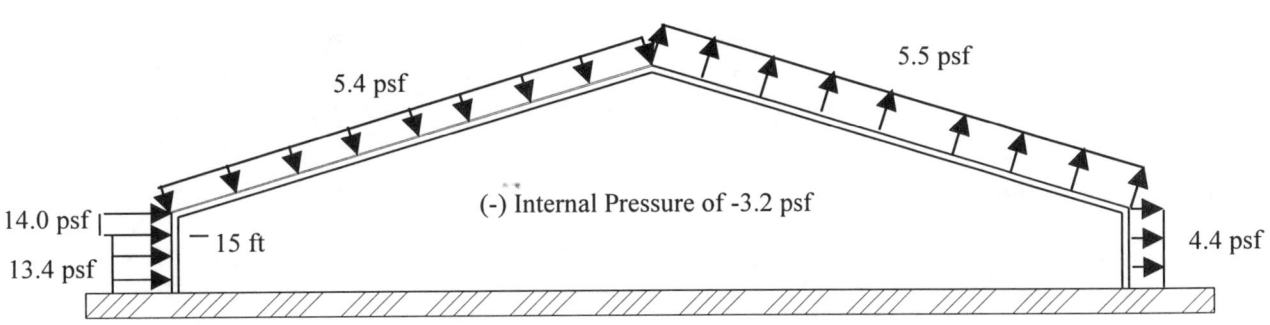

Figure 3.7.3 Net Design Wind Pressures for MWFRS when Wind is Normal to Ridge with Positive Windward External Roof Pressure Coefficient

External Roof C_p from Figure 6-3 for Wind Parallel to Ridge

For wind parallel to the ridge, $h/L = 36.7/250 = 0.147$ and $\theta < 10$ degrees. The values of C_p for wind parallel to ridge for all θ are obtained from Figure 6-3 and are shown below.

Table 3.7.5 Roof C_p (Wind Parallel to Ridge)

Surface	h/L	Distance from Windward Edge	C_p
Roof	≤ 0.5	0 to h	-0.9
		h to 2h	-0.5
		> 2h	-0.3

Table 3.7.6 MWFRS Pressures: Wind Parallel to Ridge

Surface	z ft	q psf	G	C_p	Net Pressure psf With (+GC_pi)	(-GC_pi)
Windward Wall	0-15	15.0	0.85	0.8	7.0	13.4
	20	15.8	0.85	0.8	7.5	14.0
	30	17.3	0.85	0.8	8.5	15.0
	40	18.3	0.85	0.8	9.2	15.7
	53.3	19.4	0.85	0.8	10.0	16.4
Leeward Wall	All	18.0	0.85	-0.45	-10.1	-3.7
Side Walls	All	18.0	0.85	-0.7	-14.0	-7.5
Roof*	0 to h*	18.0	0.85	-0.9	-17.0	-10.5
	h to 2h*	18.0	0.85	-0.5	-10.9	-4.4
	> 2h*	18.0	0.85	-0.3	-7.8	-1.4

Notes: $q_h = 18.0$ psf; $(GC_{pi}) = \pm 0.18$; $h = 36.7$ ft; $q_h(GC_{pi}) = \pm 3.2$ psf
* Distance from windward edge

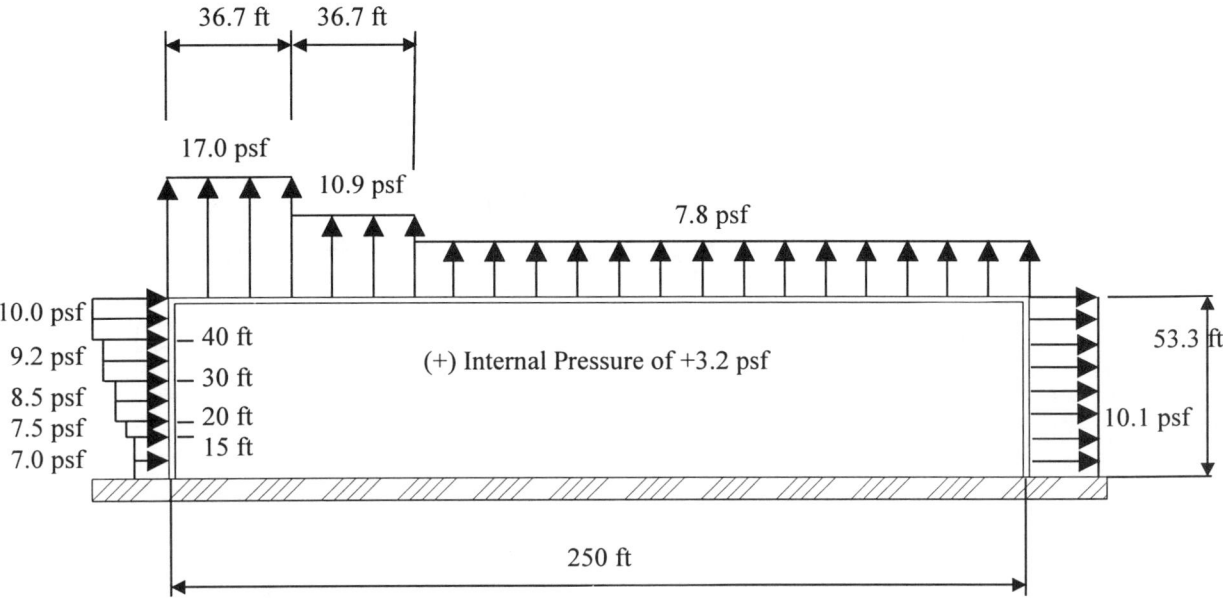

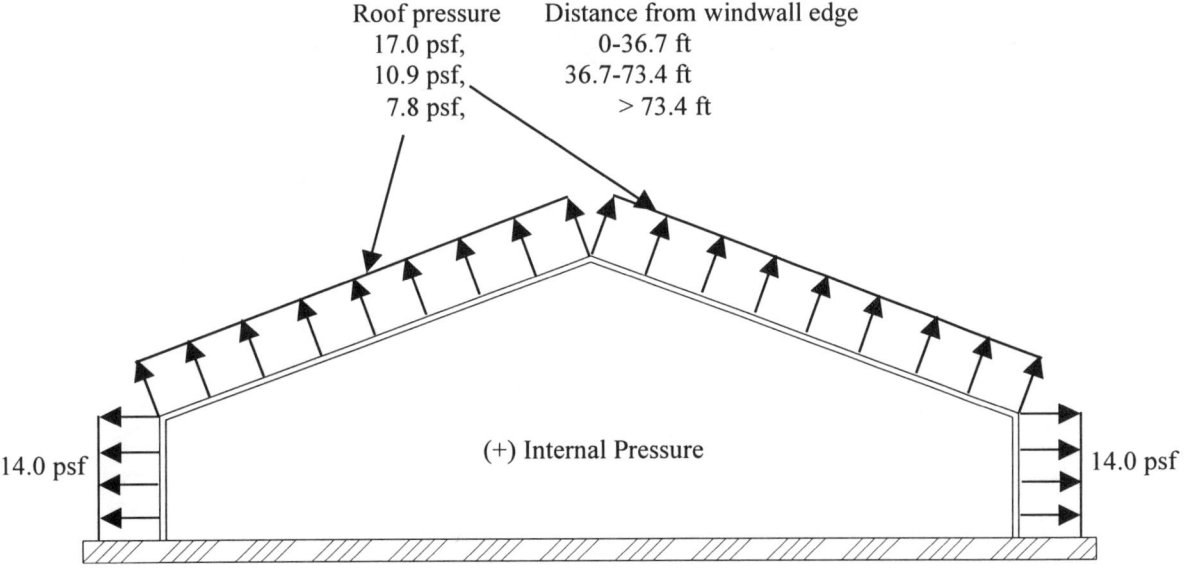

a. With Positive Internal Pressure

Figure 3.7.4 Net Design Wind Pressures for MWFRS when Wind is Parallel to Ridge

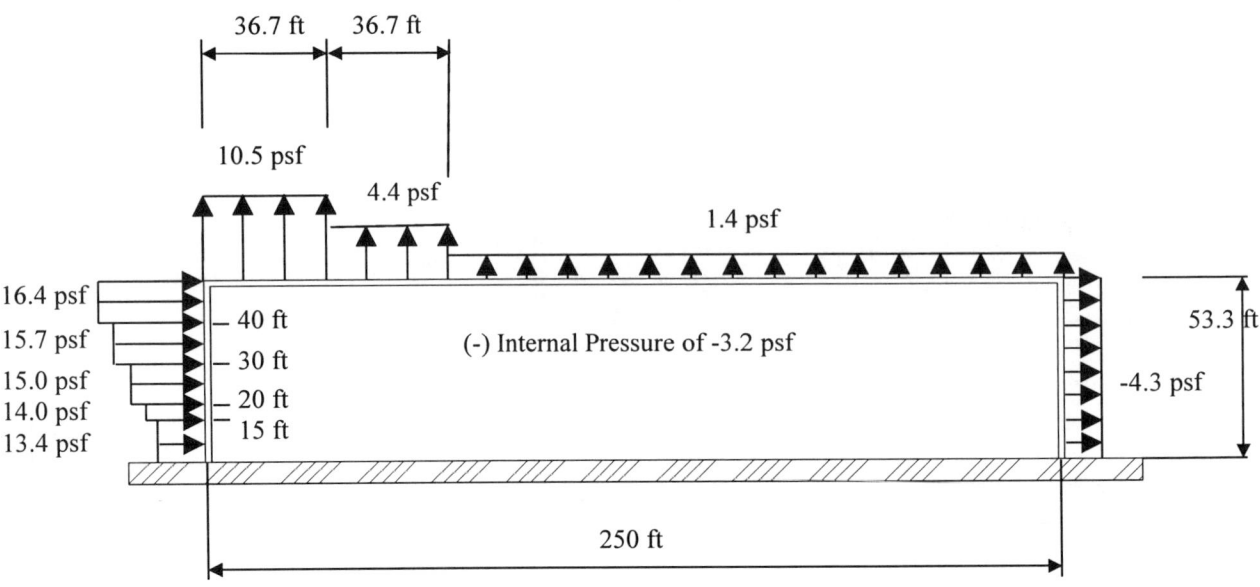

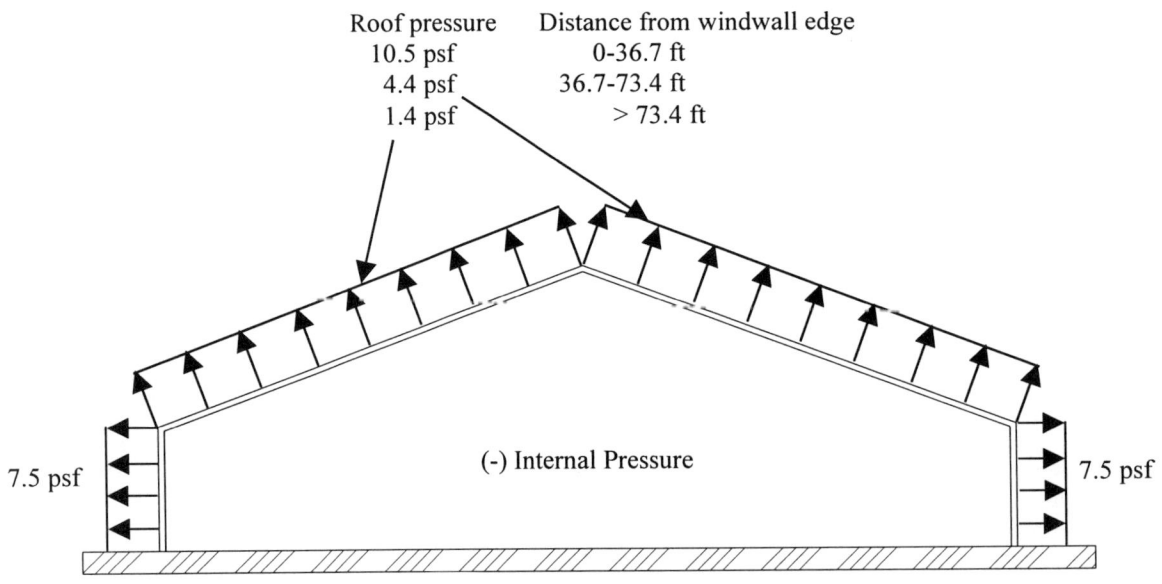

b. With Negative Internal Pressure

Figure 3.7.5 Net Design Wind Pressures for MWFRS when Wind is Parallel to Ridge

Design Pressures for Components and Cladding (C&C), Section 6.5.12.4

Equation 6-18 of the Standard is used to obtain the design pressures for components and cladding:

$$p = q_h[(GC_p) - (GC_{pi})]$$ (Equation 6-18)

where:
q_h = 18.0 psf
(GC_p) = values obtained from Figure 6-5
(GC_{pi}) = ± 0.18 for this building

Wall Components and Cladding Pressures
        The pressure coefficients (GC_p) are a function of effective wind area. The definitions of effective wind area for a component or cladding panel is the span length multiplied by an effective width that need not be less than one-third the span length; however, for a fastener it is the area tributary to an individual fastener.

Girt: larger of
                        A = 25(6.67) = 167 sq ft
        or
                        A = 25(25/3) = 208 sq ft (controls)

Wall Panel: larger of
                        A = 6.67(2) = 13.3 sq ft
        or
                        A = 6.67(6.67/3) = 14.8 sq ft (controls)

Fastener: A = 6.67(1) = 6.7 sq ft

Table 3.7.7 Wall Coefficients (GC_p) from Figure 6-5a

C&C	A sq ft	External (GC_p) Zones 4 and 5	Zone 4	Zone 5
Girt	208	0.77	-0.87	-0.93
Panel	14.8	0.97	-1.07	-1.34
Fastener	6.7	1.00	-1.10	-1.40
Other	≤ 10	1.00	-1.10	-1.40
Other	≥ 500	0.70	-0.80	-0.80

        * Other C&C can be doors, windows, etc.

Typical calculations of design pressures for Girt in Zone 4

For maximum negative pressure
$p = 18.0[(-0.87) - (\pm0.18)]$
$p = -18.9$ psf with positive internal pressure (controls)
$p = -12.4$ psf with negative internal pressure

For maximum positive pressure
$p = 18.0[(0.77) - (\pm0.18)]$
$p = 10.6$ psf with positive internal pressure
$p = 17.1$ psf with negative internal pressure (controls)

Table 3.7.8 Net Wall Component Pressures, psf

| C&C | Controlling Design Pressures, psf | | | |
| | Zone 4 | | Zone 5 | |
	Positive	Negative	Positive	Negative
Girt	17.1	-18.9	17.1	-20.0
Panel	20.7	-22.5	20.7	-27.4
Fastener	21.2	-23.0	21.2	-28.4
$A \leq 10$ sq ft	21.2	-23.0	21.2	-28.4
$A \geq 500$ sq ft	15.8	-17.6	15.8	-17.6

Roof Components and Cladding Pressures
Effective wind areas of roof C&C

Purlin: larger of
                    $A = 25(5) = 125$ sq ft
          or $A = 25(25/3) = 208$ sq ft (controls)

Panel: larger of
                    $A = 5(2) = 10$ sq ft (controls)
          or $A = 5(5/3) = 8.3$ sq ft

Fastener: $A = 5(1) = 5$ sq ft

Table 3.7.9 Roof Coefficients (GC_p) from Figure 6-5b; $10 < \theta \leq 30°$

| Components | A sq ft | External (GC_p) | | |
		Zones 1, 2, and 3	Zone 1	Zones 2 and 3
Purlin	208	0.3	-0.8	-1.4
Panel	10	0.5	-0.9	-2.1
Fastener	5	0.5	-0.9	-2.1
Other	≤ 10	0.5	-0.9	-2.1
Other	≥ 100	0.3	-0.8	-1.4

* Other C&C can be skylight, etc.

Typical calculations of design pressures for purlin in Zone 1

For maximum negative pressure
$p = 18.0[(-0.8) - (\pm 0.18)]$
$p = -17.6$ psf with positive internal pressure (controls)
$p = -11.2$ psf with negative internal pressure

For maximum positive pressure
$p = 18.0[(0.3) - (\pm 0.18)]$
$p = 2.1$ psf with positive internal pressure
$p = 8.6$ psf with negative internal pressure
$p = 10$ psf minimum net pressure (controls); 6.1.4.2

Table 3.7.10 Net Roof Component Pressures, psf

| Component | Controlling Design Pressures, psf | | |
| | Positive | Negative | |
	Zones 1, 2, and 3	Zone 1	Zones 2 and 3
Purlin	10.0*	-17.6	-28.4
Panel	12.2	-19.4	-41.0
Fastener	12.2	-19.4	-41.0
$A \leq 10$ sq ft	12.2	-19.4	-41.0
$A \geq 500$ sq ft	10.0*	-17.6	-28.4

* minimum net pressure controls, 6.1.4.2

Special case of girt that transverses Zones 4 and 5

Width of Zone 5:
smaller of
$\quad a = 0.1(200) = 20$ ft
or
$\quad a = 0.4(36.7) = 14.7$ ft (controls)
but not less than $0.04(200) = 8$ ft
$\quad$ or 3 ft

Weighted average design pressure
$$P = \frac{14.7(-20.0) + 10.3(-18.9)}{25} = -19.6 \text{ psf}$$

This procedure of using weighted average may be used for other components and cladding.

Special Case of Strut Purlin
    Strut purlins in the end bay experience combined uplift pressure as a roof component (C&C) and axial load as part of the MWFRS

- Component Pressure
    Internal purlin is located in Zones 1 and 2
    Width of Zone 2, a = 14.7 ft
    Weighted average design pressure

$$= \frac{14.7(-28.4) + 10.3(-17.6)}{25} = -24.0 \text{ psf}$$

- MWFRS Load
    Figure 3.7.4 shows design pressure on end wall with wind parallel to ridge with positive internal pressure (consistent with high uplift on the purlin). Assuming that the end wall is supported at the bottom and at the roof line, the effective axial load on an interior purlin can be determined.

- Combined design load on interior strut purlin

24.0 psf (calculated above)

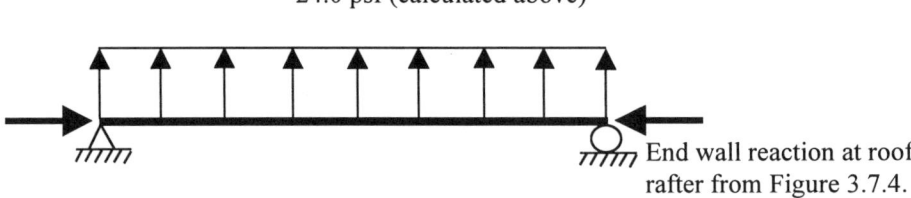

End wall reaction at roof
rafter from Figure 3.7.4.

    Note that many metal building manufacturers support the top of the wall panels with the eave strut purlin, see Figure 3.7.6. For this case, the eave purlin also serves as a girt and the negative wall pressures of Zones 5 and 4 would occur for the same wind direction as the maximum negative uplift pressures on the purlin, refer to Zones 3 and 2. Thus in this instance, the correct load combination would involve bi-axial bending loads based on component and cladding pressures combined with the MWFRS axial load.

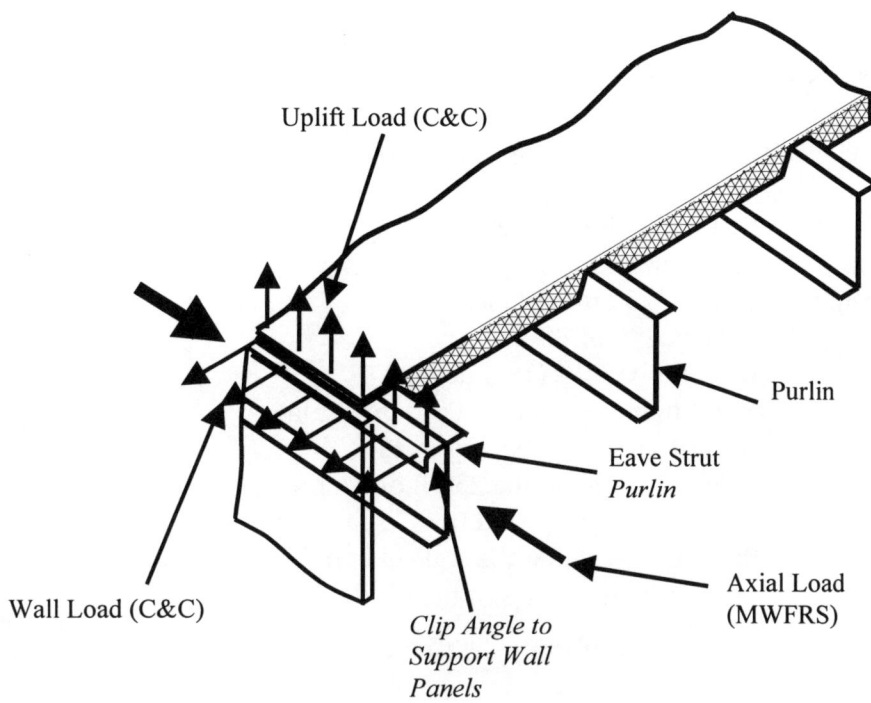

Figure 3.7.6 Eave Strut Purlin Supports Roof and Wall Panels

3.8 Example 8 – Building of Example 7 Using Height h < 60 ft and Low-Rise Building Provisions

This example illustrates the use of the low-rise building provisions using Figure 6-4 of ASCE 7-98 to determine design pressures for the MWFRS. For this purpose, the building used has the same dimensions as Example 7. The design pressures on components and cladding will be the same as Example 7. The building data are as follows:

Location: Memphis, Tennessee
Terrain: Flat farmland
Dimensions: 200 ft x 250 ft in plan
           Eave height of 20 ft
           Roof Slope 4:12 (18.4 degrees)
Framing: Rigid frame spans the 200 ft direction
           Rigid frame bay spacing is 25 ft
           Lateral bracing in the 250 ft direction
           is provided by a "wind truss" spanning
           the 200 ft to side walls and cable/rod
           bracing in the planes of the walls

           Openings uniformly distributed

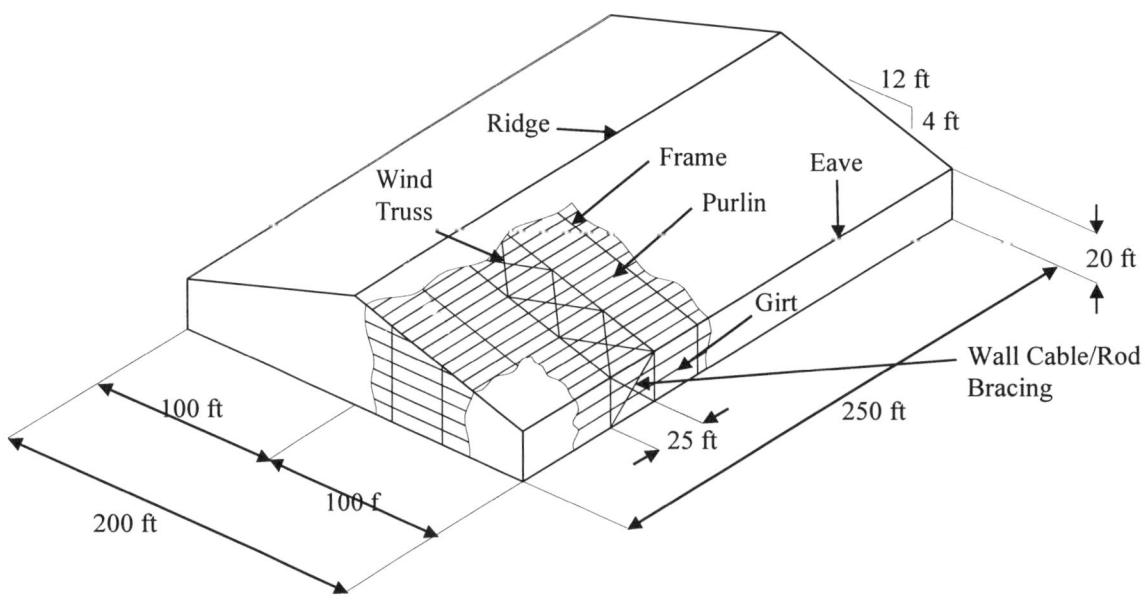

Figure 3.8.1 Dimensions and Framing of the Low-Rise Building of Example 8

Low-Rise Building

Section 6.2 of the Standard specifies two requirements for a building to qualify as a low-rise building: (1) mean roof height has to be less or equal to 60 ft, and (2) mean roof height does not exceed least horizontal dimension. A building with these dimensions qualifies as a low-rise building and the external pressure coefficients of Figure 6-4 may be used.

Exposure, Building Classification, and Basic Wind Speed

Same as Example 6.1:   Exposure C
                            Category II
                            Enclosed Building (openings uniformly distributed)
                            $V = 90$ mph

Velocity Pressure

The low-rise building provisions for MWFRS in the Standard use the velocity pressure at mean roof height h for calculation of all external and internal pressures including the windward wall. All pressures for a given zone are assumed to be uniformly distributed with respect to height above ground.

Mean roof height $h = 36.7$ ft
The velocity pressures are computed using:

$$q_h = 0.00256 K_h K_{zt} K_d V^2 I \text{ (psf)} \qquad \text{(Equation 6-1)}$$

where:
q_h = velocity pressure at mean roof height h
$K_h = 1.02$ for Exposure C, see Table 6-5 Case 1
$K_{zt} = 1.0$ topographic factor, see Section 6.5.7.1
$K_d = 0.85$, see Table 6-6
$V = 90$ mph basic wind speed, see Figure 6-1
$I = 1.0$ for Category II (50-year mean return interval)
therefore,
$q_h = 0.00256(1.02)(1.0)(0.85)(90)^2(1.0) = 18.0$ psf

Design Pressures for the MWFRS

The equation for the determination of design wind pressures for MWFRS for low-rise buildings is given by equation 6-16 in Section 6.5.12.2.2:

$$p = q_h[(GC_{pf}) - (GC_{pi})]$$ (Equation 6-16)

where:

q_h is the velocity pressure at mean roof height associated with Exposure C
(GC_{pf}) is the external pressure coefficients from Figure 6-4
(GC_{pi}) is the internal pressure coefficient from Table 6-7

To obtain the critical wind actions, including torsion, for design of the MWFRS, the two separate loading cases, Case A and Case B, indicated in Figure 6-4 must be applied at each corner, see Figure C6-2. For each of these cases, both positive and negative internal pressures must be considered, resulting in a total of 16 separate loading conditions. However, if the building is symmetrical, the number of separate loading conditions will be reduced to eight (two loading cases, two windward corners, and two internal pressures).

In applying the loads of Case A and Case B to corner 2 shown in Figure 6.4 for the building rotated 90 degrees, see Figure C6-2, Zones 2 and 3 are assumed to be separated by an imaginary ridge line oriented normal to and bisecting the actual ridge line. Additionally, the roof angle, θ, is assumed to be zero degrees when selecting the values of GC_{pf} for Case A loading for wind applied at corner 2.

External Pressure Coefficients (GC$_{pf}$)

The roof and wall coefficients are functions of the roof slope, θ. There are eight building surfaces defined for Case A: four interior surfaces and four end zone surfaces. For Case B, there are 12 surfaces identified: six interior surfaces and six end zone surfaces.

Width of end zone surface: smaller of
$2a = 2(0.1)(200) = 40$ ft
or $2(0.4)(36.7) = 29.4$ ft (controls)
but not less than $2(0.04)(200) = 16$ ft
or $2(3) = 6$ ft

Table 3.8.1 Corner 1: Case A, GC$_{pf}$ ($\theta = 18.4°$)

$\theta°$	Building Surface							
	1	2	3	4	1E	2E	3E	4E
0-5	0.40	-0.69	-0.37	-0.29	0.61	-1.07	-0.53	-0.43
18.4*	0.52	-0.69	-0.47	-0.42	0.78	-1.07	-0.67	-0.62
20	0.53	-0.69	-0.48	-0.43	0.80	-1.07	-0.69	-0.64

* By linear interpolation

Table 3.8.2 Corner 1 and 2: Case B, GC_{pf} (all roof angles)

Zone	Building Surface					
Location	1	2	3	4	5	6
Interior	-0.45	-0.69	-0.37	-0.45	0.40	-0.29
Edge*	-0.48	-1.07	-0.53	-0.48	0.61	-0.43

* Zones 1E, 2E, 3E, 4E, 5E, and 6E in Figure 6-4

Table 3.8.3 Corner 2: Case A, $GC_{pf}(\theta = 0°)$

$\theta°$	Building Surface							
	1	2	3	4	1E	2E	3E	4E
0-5	0.40	-0.69	-0.37	-0.29	0.61	-1.07	-0.53	-0.43

Internal Pressure Coefficients (GC$_{pi}$)

Openings are assumed to be evenly distributed in the walls and since Memphis, Tennessee is not located in a hurricane-prone region, the building qualifies as an enclosed building, see Section 6.2 and the internal pressure coefficients are given from Table 6-7 as

$$(GC_{pi}) = \pm0.18$$

Design Wind Pressures, psf

Typical calculations for design pressures are as follows:

Corner 1, Case A: Surface 1
  $p = 18.0[(0.52) - (\pm0.18)]$
  $p = 12.6$ psf with (-) internal pressure
  $p = 6.1$ psf with (+) internal pressure
Corner 1, Case A: Surface 2E
  $p = 21.2[(-1.07) - (\pm0.18)]$
  $p = -16.0$ psf with (-) internal pressure
  $p = -22.5$ psf with (+) internal pressure

Table 3.8.4 Design Wind Pressures, Corner 1: Case A

Building Surface	(GC$_{pf}$)	Design Pressure, psf	
		(+GC$_{pi}$)	(-GC$_{pi}$)
1	0.52	6.1	12.6
2	-0.69	-15.6	-9.2
3	-0.47	-11.7	-5.2
4	-0.42	-10.8	-4.3
1E	-0.78	10.8	17.3
2E	-1.07	-22.5	-16.0
3E	-0.67	-15.3	-8.8
4E	-0.62	-14.4	-7.9

Table 3.8.5 Design Wind Pressures, Corner 1 and 2: Case B

Building Surface	(GC$_{pf}$)	Design Pressure, psf	
		(+GC$_{pi}$)	(-GC$_{pi}$)
1	-0.45	-11.3	-4.9
2	-0.69	-15.6	-9.2
3	-0.37	-9.9	-3.4
4	-0.45	-11.3	-4.9
5	0.40	4.0	10.5
6	-0.29	-8.4	-2.0
1E	-0.48	-11.9	-5.4
2E	-1.07	-22.5	-16.0
3E	-0.53	-12.8	-6.3
4E	-0.48	-11.9	-5.4
5E	0.61	7.7	14.2
6E	-0.43	-11.0	-4.5

Table 3.8.6 Design Wind Pressures, Corner 2: Case A

Building Surface	(GC$_{pf}$)	Design Pressure, psf	
		(+GC$_{pi}$)	(-GC$_{pi}$)
1	0.40	4.0	10.5
2	-0.69	-15.6	-9.2
3	-0.37	-9.9	-3.4
4	-0.29	-7.6 ~~8.5~~	-2.0
1E	0.61	7.7	14.2
2E	-1.07	-22.5	-16.0
3E	-0.53	-12.8	-6.3
4E	-0.43	-11.0	-4.5

Application of Pressures on Building Surfaces 2 and 3

Footnote 4(a) of Figure 6-4 states that the roof pressure coefficient GC_{pf}, when negative in Zone 2, shall be applied in Zone 2 for a distance from the edge of the roof equal to 0.5 times the horizontal dimension of the building measured perpendicular to the eave line or 2.5h, whichever is less; the remainder of Zone 2 extending to the ridge line shall use the pressure coefficient GC_{pf} for Zone 3. Thus, the distance from the edge of the roof is the smaller of:

$$0.5(200) = 100 \text{ ft}$$
$$\text{or}$$
$$(2.5)(36.7) = 92 \text{ ft (controls)}$$

Therefore, Zone 3 applies over a distance of $105 - 92 = 13$ ft in what is normally considered to be Zone 2 (adjacent to ridge line).

Loading Cases

Because the building is symmetrical, the eight loading cases provide all the required combinations provided the design is accomplished by applying loads for each of the four corners. The load combinations illustrated in Figures 3.8.2 through 3.8.9 are to be used to design the rigid frames, the "wind truss" spanning across the building in the 200 ft direction, and the rod/cable bracing in the planes of the walls, see Figure 3.8.1.

Design Wind Pressures for Components and Cladding

The design pressures for components and cladding are the same as shown in Example 7.

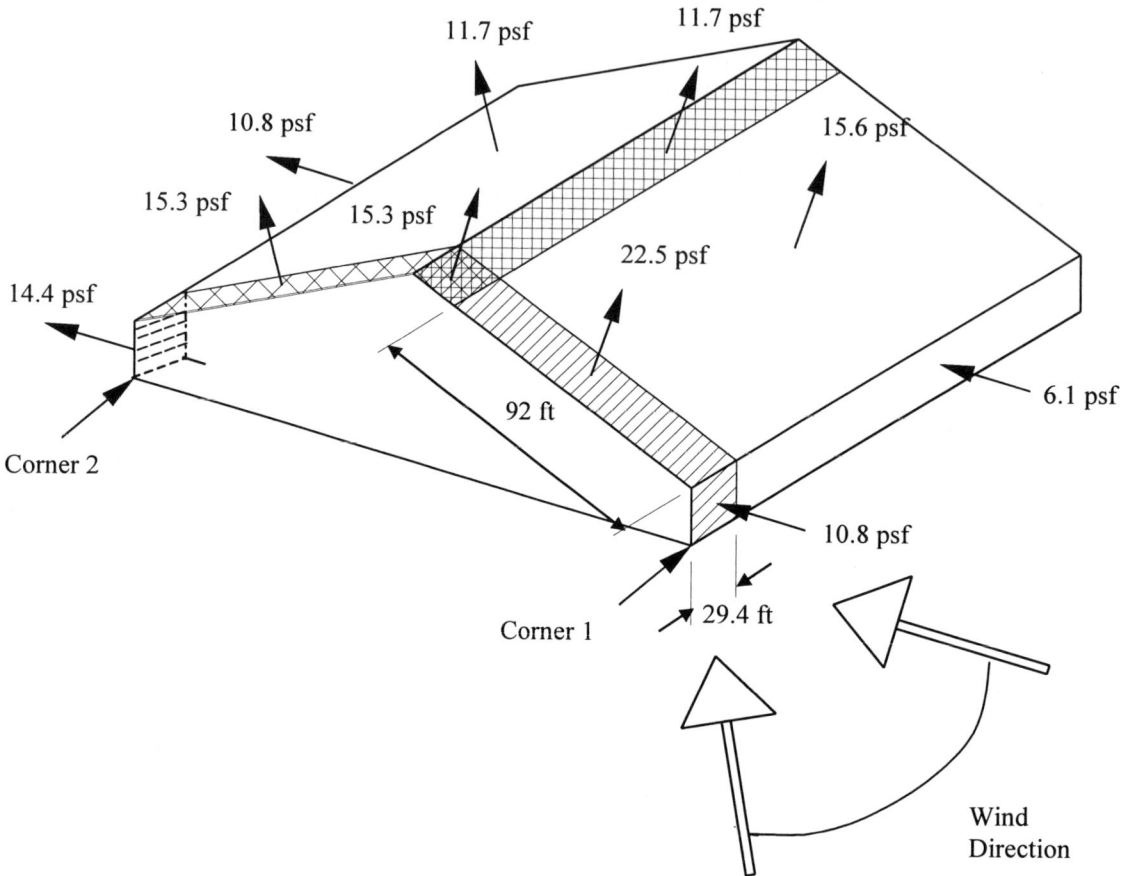

Notes:
1. The pressures are assumed to be uniformly distributed over each of the surfaces shown
2. Roof pressures of 15.6 and 22.5 psf apply up to 92 ft; the remaining 13 ft up to the ridge line will have pressures of 11.7 and 15.3 psf

Figure 3.8.2 Design Pressures for Case A at Corner 1 with Positive Internal Pressure

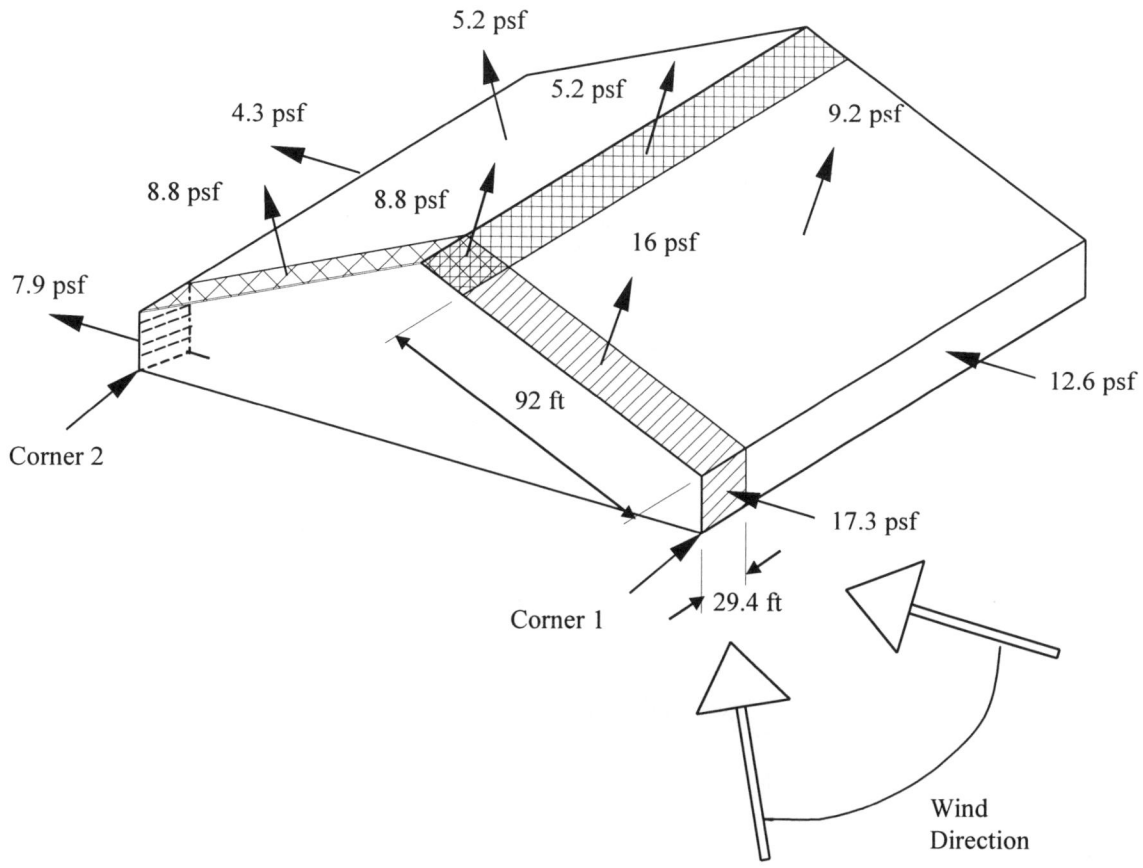

Notes:

1. The pressures are assumed to be uniformly distributed over each of the surfaces shown
2. Roof pressures of 9.2 and 16 psf apply up to 92 ft; the remaining 13 ft up to the ridge line will have pressures of 5.2 and 8.8 psf

Figure 3.8.3 Design Pressures for Case A at Corner 1 with Negative Internal Pressure

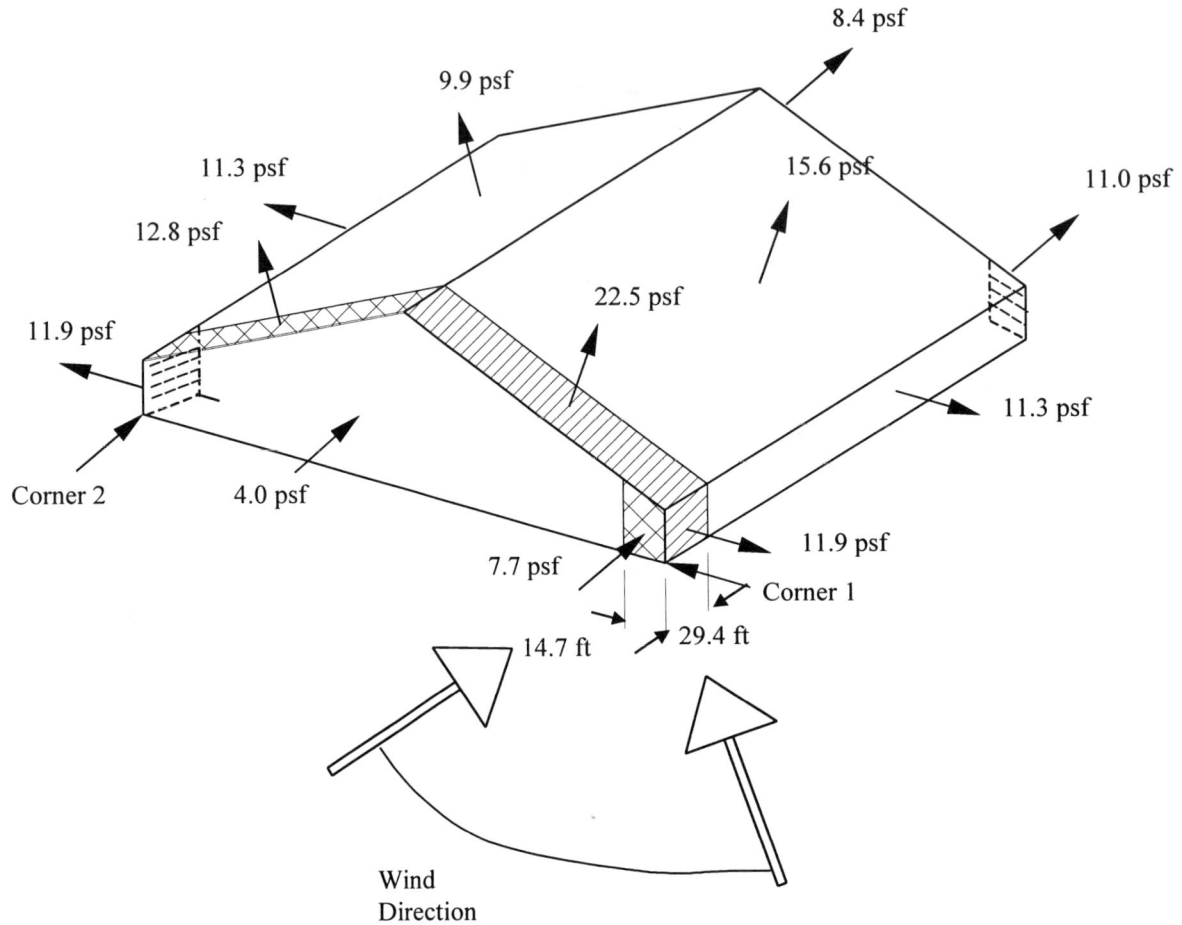

Notes:
1. The pressures are assumed to be uniformly distributed over each of the surfaces shown

Figure 3.8.4 Design Pressures for Case B at Corner 1 with Positive Internal Pressure

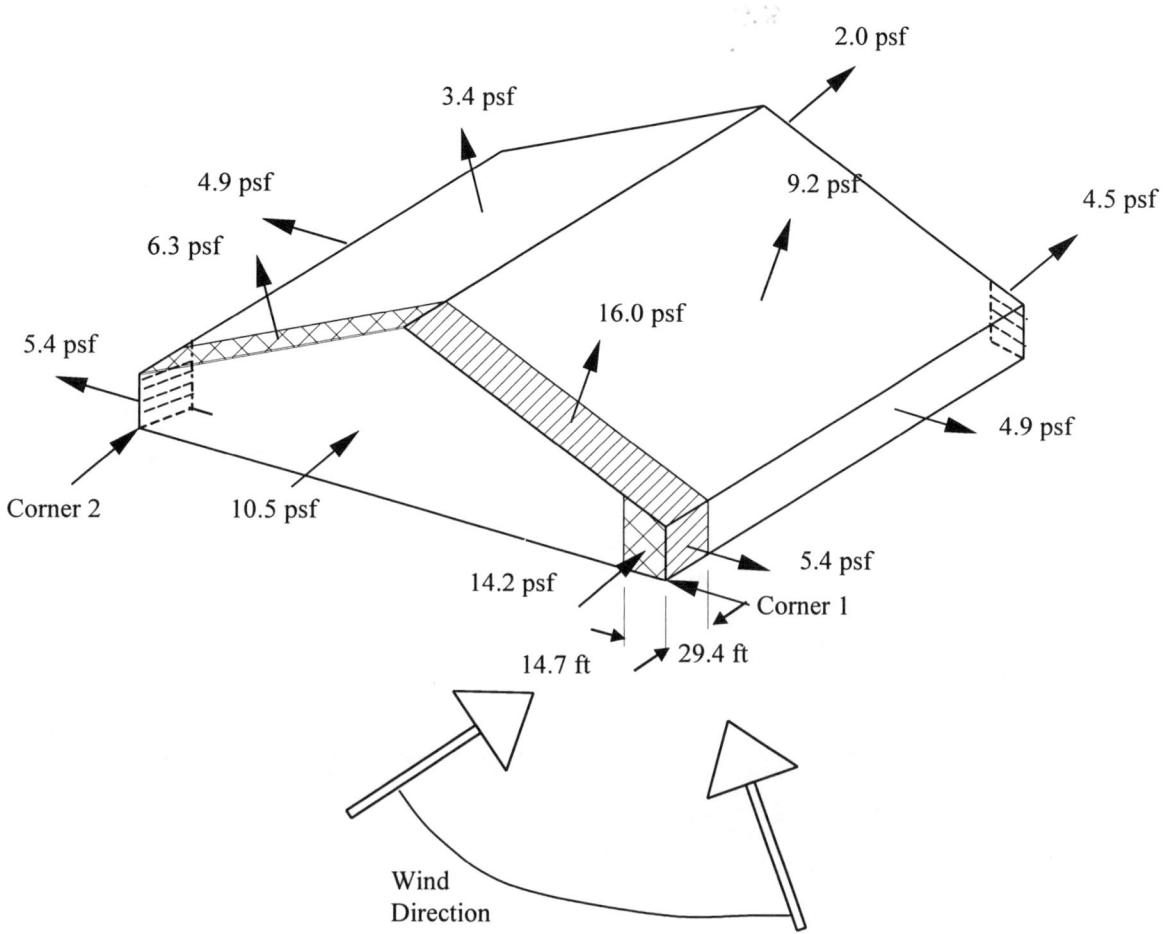

Notes:
1. The pressures are assumed to be uniformly distributed over each of the surfaces shown

Figure 3.8.5 Design Pressures for Case B at Corner 1 with Negative Internal Pressure

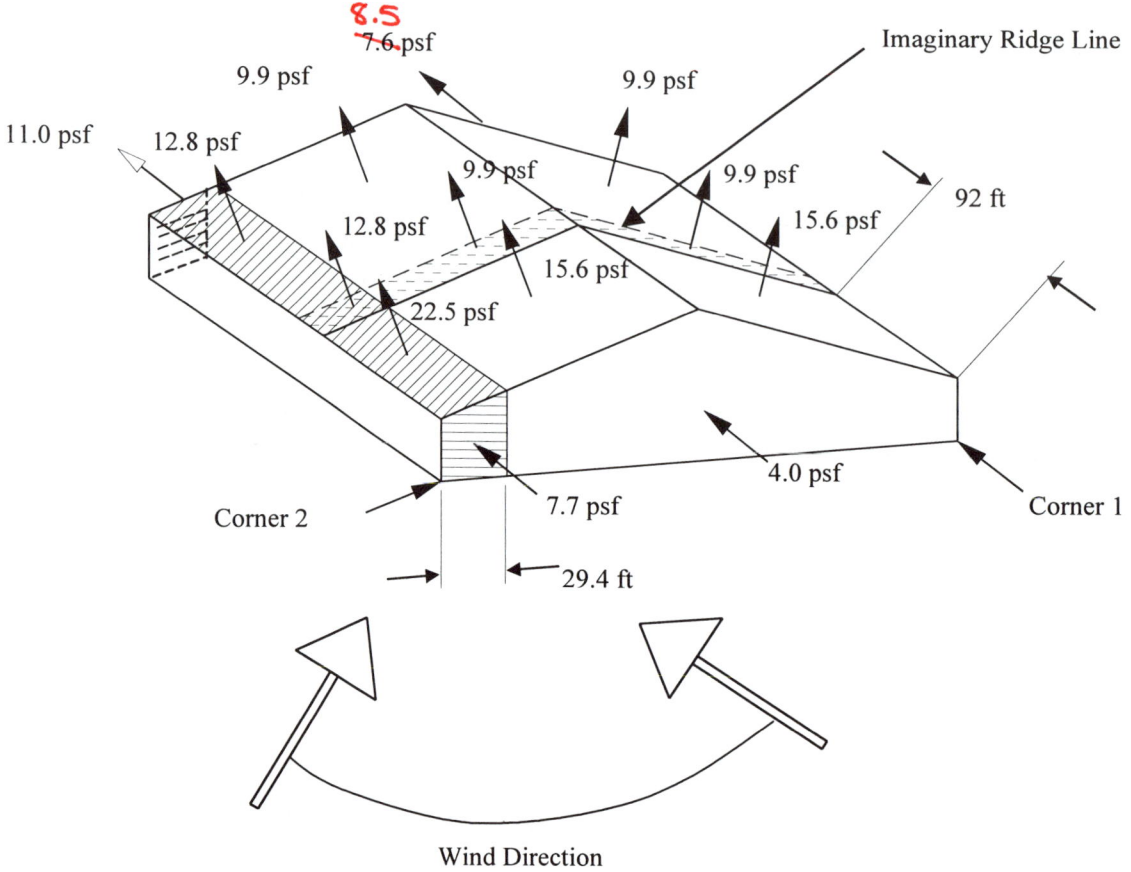

Figure 3.8.6 Design Pressures for Case A at Corner 2 with Positive Internal Pressure

Notes:
1. The pressures are assumed to be uniformly distributed over each of the surfaces shown
2. Roof pressures of 15.6 and 22.5 psf apply up to 92 ft; the remaining 33 ft up to the ridge line will have pressures of 9.9 and 12.8 psf

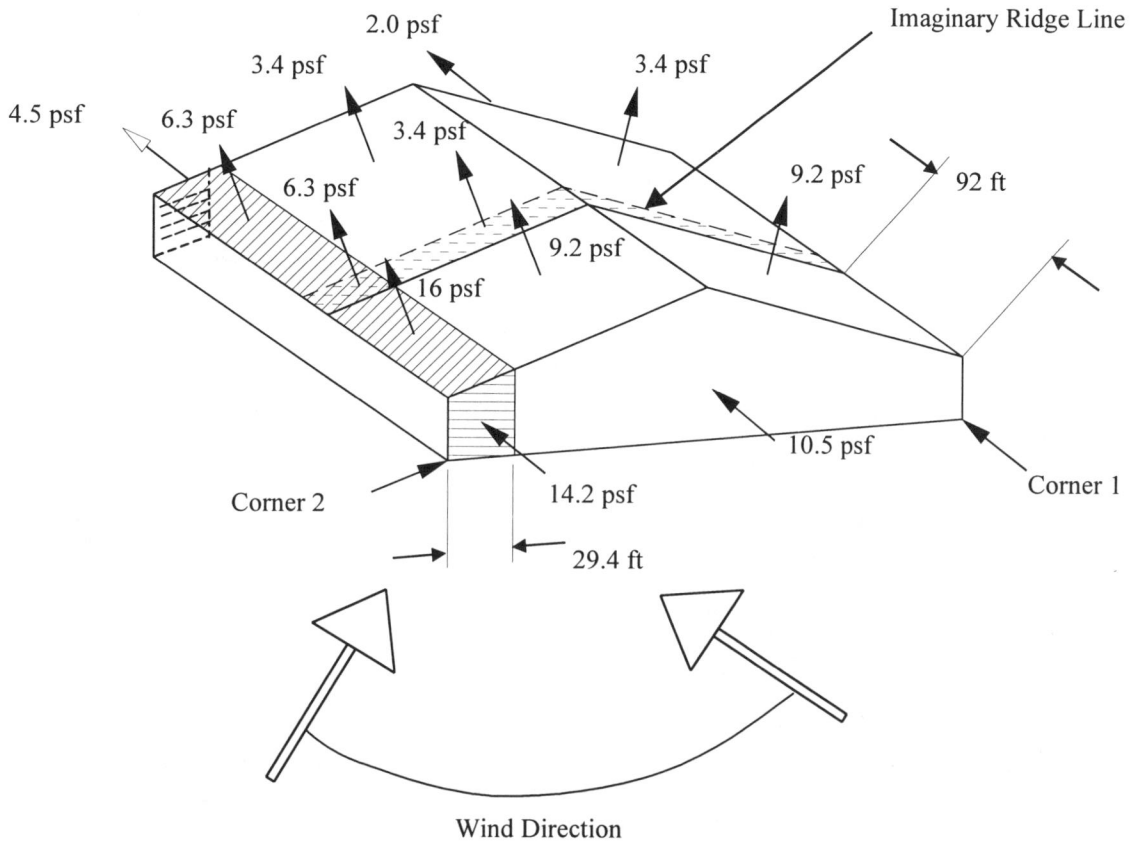

Notes:
1. The pressures are assumed to be uniformly distributed over each of the surfaces shown
2. Roof pressures of 9.2 and 16 psf apply up to 92 ft; the remaining 33 ft up to the ridge line will have pressures of 3.4 and 6.3 psf

Figure 3.8.7 Design Pressures for Case A at Corner 2 with Negative Internal Pressure

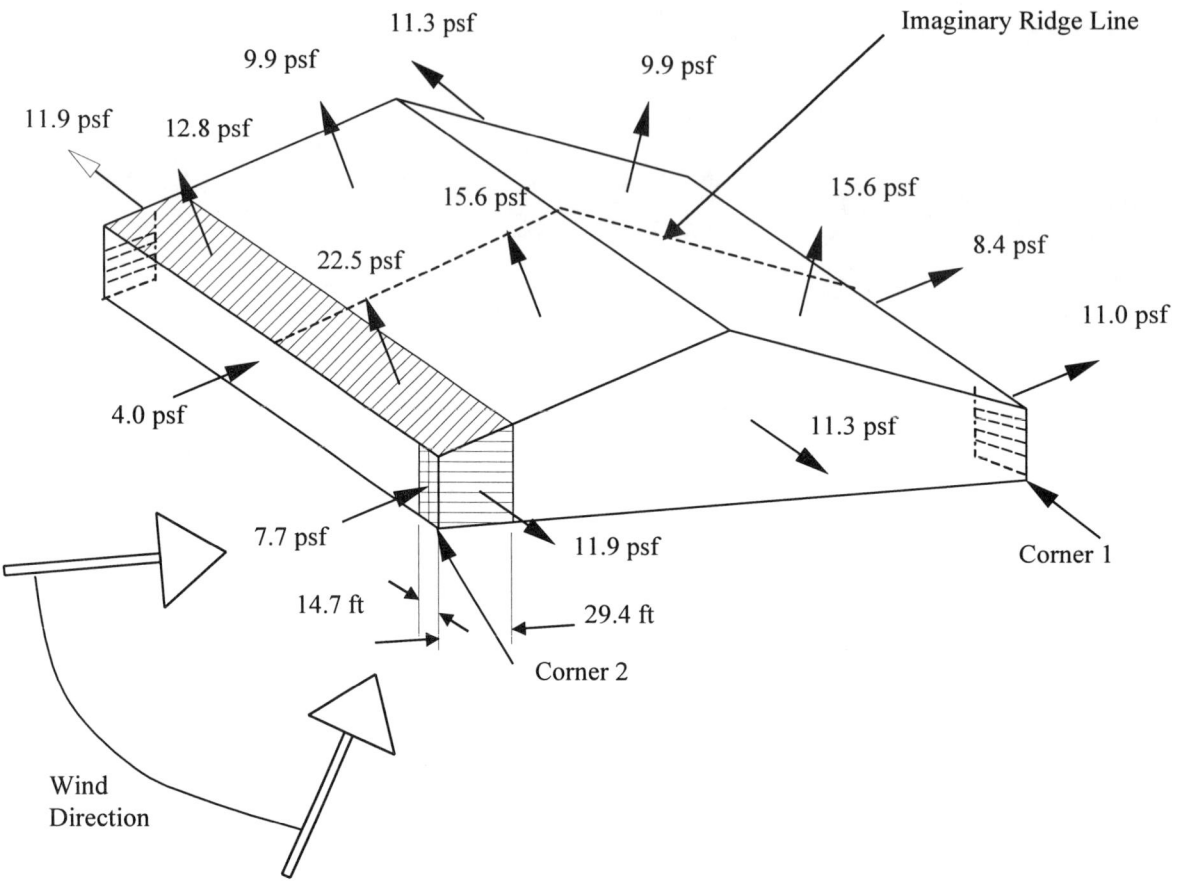

Notes:
1. The pressures are assumed to be uniformly distributed over each of the surfaces shown

Figure 3.8.8 Design Pressures for Case B at Corner 2 with Positive Internal Pressure

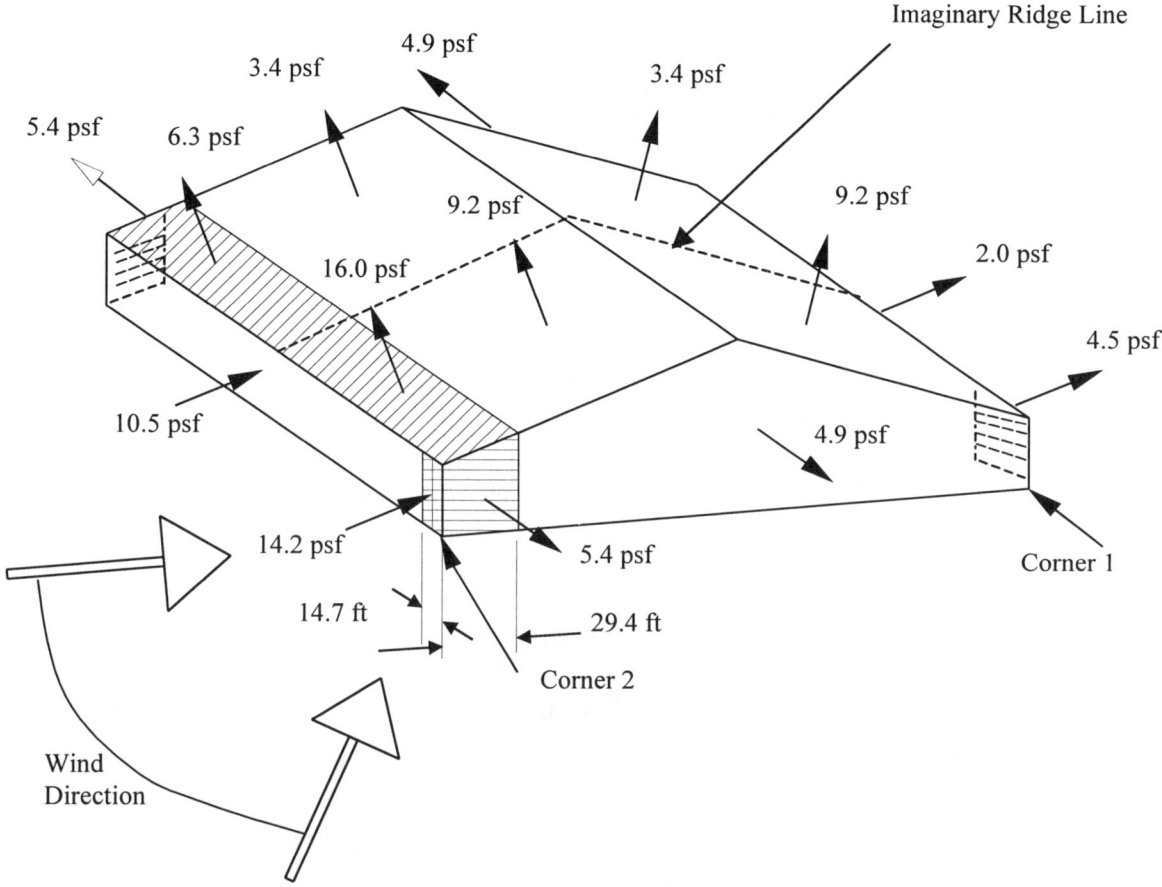

Notes:

1. The pressures are assumed to be uniformly distributed over each of the surfaces shown

Figure 3.8.9 Design Pressures for Case B at Corner 2 with Negative Internal Pressure

Comment: Exposure B Terrain

If the building were to be sited in terrain representative of Exposure B in <u>all</u> directions, the low-rise MWFRS and C&C design pressures would be calculated using the velocity pressure coefficients for Case 1, Exposure B of Table 6-5. Thus for h = 36.7 ft, $K_{36.7}$ = 0.74. The velocity pressure at mean roof height would be given by

$$q_h = (0.00256)(0.74)(1.0)(0.85)(90)^2(1.0) = 13 \text{ psf}$$

resulting in MWFRS loads of 13/18 = 72% of those given in Figures 3.9.2 through 3.9.9.

3.9   Example 9 – 40 ft x 80 ft Commercial Building with Monoslope Roof with Overhang

In this example, design pressures for a typical retail store in a strip-mall are determined. The building data are as follows:

Location:	Boston, Massachusetts within one mile of the coastal mean high water mark
Topography:	Homogeneous
Terrain:	Suburban
Dimensions:	40 ft x 80 ft in plan
	Monoslope roof with slope of 14 degrees and overhang of 7 ft in plan
	Wall heights are 15 ft in front and 25 ft in rear
Framing:	Walls of CMU on all sides supported at top and bottom; steel framing in front (80 ft width) to support window glass and doors. Roof joists span 41.2 ft with 7.2 ft overhang spaced at 5 ft on center
Cladding:	Glass and door sizes vary; glazing is not debris impact resistant and occupies 50% of front wall (80 ft in width). Roof panels are 2 ft wide and 20 ft long

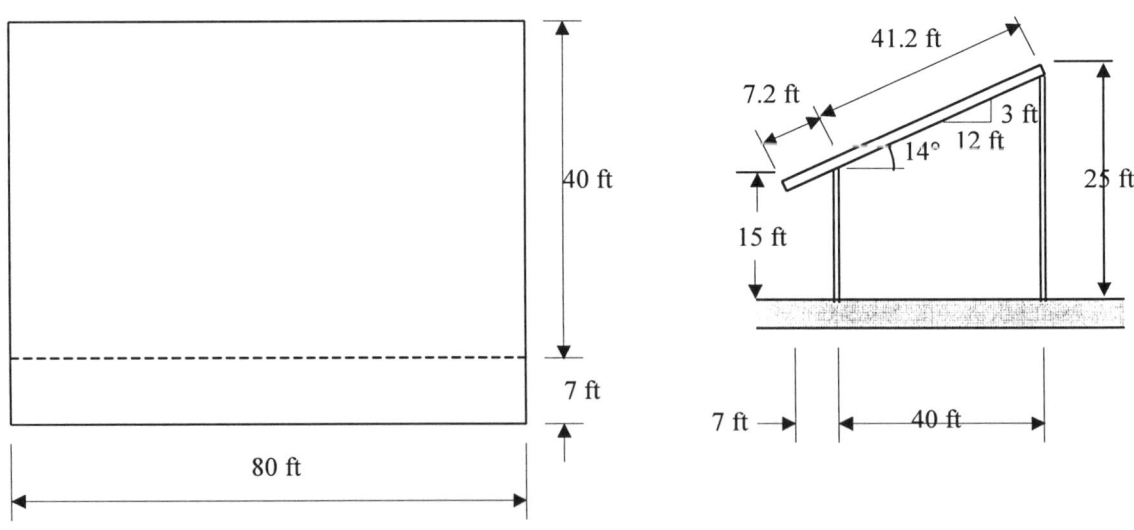

Figure 3.9.1 Dimensions of the Retail Store Strip-Mall

Building Classification, Enclosure Classification, and Exposure Category

The building is not an essential facility nor likely to be occupied by more than 300 persons at any one time. Use Category II, see Table 1-1. Importance Factor I = 1.00, see Table 6-1.

The building is sited in a suburban area and satisfies the criteria for Exposure B, see Section 6.5.6.

The building is sited in a wind-borne debris region, has glazing not impact resistant occupying 50% of a wall receiving positive pressure; building must be classified as partially enclosed, see Sections 6.5.9.3 and 6.2.

The building does not meet the requirements of Section 6.4 Method 1 - Simplified Procedure, roof slope greater than 10 degrees; use analytical procedure of Method 2, Section 6.5.3. Roof is not gabled; hence, low-rise building provisions may not be used.

The provisions of the standard do not permit use of external pressure coefficients (GC_{pf}) given in Figure 6-4; the values in Figure 6-4 were obtained from wind tunnel studies of rigid, gable-framed buildings. Their use for a monoslope roof requires considerable judgment. The design examples presented in Example 8 illustrates use of the pressure coefficients of Figure 6-4, and the commentary in the standard gives the background for (GC_{pf}) values.

Basic Wind Speed

The wind speed contour of 110 mph traverses over Boston, Massachusetts, see Figure 6-1c; use a basic wind speed of 110 mph.

Velocity Pressures Section 6.5.10

The velocity pressures are calculated using:

$$q = 0.00256K_zK_{zt}K_dV^2I \text{ (psf)} \qquad \text{(Equation 6-13)}$$
$$= 0.00256K_z(1.0)(0.85)(110)^2(1.0)$$
$$= 26.33K_z \text{ (psf)}$$

where,
K_z is obtained from Table 6-5
K_{zt} = 1.0 homogeneous terrain
I = 1.0 for Category II building, see Table 6-1
K_d = 0.85, see Table 6-6

The provisions of the Standard require the use of the external pressure coefficients C_p from Figure 6-3; hence, the exposure coefficients K_z are based on Exposure B, Case 2 for MWFRS and Exposure B, Case 1 for C&C, see Table 6-5.

Table 3.9.1 Velocity Pressures, q_z, q_i, and q_h, psf

Height, ft	Main Force Resisting System		Component and Cladding	
	Exposure B, Case 2	q_z, q_i	Exposure B, Case 1	q_h
0-15	0.57	15.01		
$h = 20$	0.62	16.32	0.70	18.43
25	0.66	17.38		

Design Pressures for Main Wind-Force Resisting System (MWFRS)

The equation for rigid buildings of all heights is given in Section 6.15.12.2 as:

$$p = qGC_p - q_i(GC_{pi}) \qquad \text{(Equation 6-15)}$$

Where q is q_z for windward wall and $q_i = q_h$ for windward and leeward walls, side walls and roof. G is determined from Section 6.5.8. Values of C_p are obtained from Figure 6-3 and values of (GC_{pi}) are obtained from Table 6-7. For positive internal pressure evaluation, the standard permits q_i to be conservatively evaluated at height h ($q_i = q_h$). Section 6.5.11.1.1 permits the reduction of GC_{pi} for a partially enclosed building containing a single, unpartitioned large volume by the factor:

$$R_i = 0.5(1 + \frac{1}{\sqrt{1 + \dfrac{V_i}{22,800 A_{og}}}}) = 0.5(1 + \frac{1}{\sqrt{1 + \dfrac{80 \times 40 \times 20}{22,800 \times 50\%(15 \times 80)}}}) \cong 1 \text{ (no reduction)}$$

where, V_i = unpartitioned internal volume and A_{og} = total area of openings in building envelope (50% of front wall).

Note:

The provisions of the Standard do not permit the use of external pressure coefficients (GC_{pf}) given in Figure 6-4; the values in Figure 6-4 were obtained from wind tunnel studies of rigid, gable-framed buildings. Their use for a monoslope roof is not appropriate. The design in Example 8 illustrates the use of the pressure coefficients of Figure 6-4, and the commentary in the Standard gives the background for (GC_{pf}) values.

Gust Effect Factor, G

The gust effect factor for non-flexible (rigid) buildings is given in Section 6.5.8 as:

$$G = 0.85$$

The size of the building would not permit a reduction in G based on Equation 6-2.

Wall External Pressure Coefficients, C_p

The coefficients for the windward and side walls are given in Figure 6-3 as C_p = +0.8 and -0.7, respectively. The values for the leeward wall depend on L/B; they are different for the two directions: (1) wind parallel to roof slope (normal to ridge), and (2) wind normal to roof slope (parallel to ridge).

Table 3.9.2 Wall Pressure Coefficients, C_P

Surface	Wind Direction	L/B	C_P
Leeward Wall	‖ to roof slope	0.5	-0.5
Leeward Wall	⊥ to roof slope	2.0	-0.3
Windward Wall	-	-	0.8
Side Walls	-	-	-0.7

Roof External Pressure Coefficients, C_p

Since the building has a monoslope roof, the roof surface for wind directed parallel to the slope (normal to ridge) may be a windward or a leeward surface. The value of h/L = 0.5 in this case, and the proper coefficients are obtained from linear interpolation for θ = 14 degrees.

When wind is normal to the roof slope (parallel to ridge), angle θ = 0 and h/L = 0.25.

Table 3.9.3 Roof Pressure Coefficients, C_p

Wind Direction	h/L	θ°	C_p
‖ to roof slope	0.5	14	-0.74 as windward slope
‖ to roof slope	0.5	14	-0.50 as leeward slope
⊥ to roof slope	0.25	0	-0.90 (0-20 ft)* -0.50 (20-40 ft) -0.30 (40-80 ft)

* Distance from the windward edge of the roof

For the overhang, Section 6.5.11.4.1 requires $C_p = 0.8$ for wind directed normal to 15 ft wall. The standard is silent on the leeward overhang for the case of wind directed toward 25 ft wall and perpendicular to roof slope (parallel to ridge). One could use $C_p = -0.5$ (coefficient for leeward wall) but coefficient has been conservatively taken as zero.

The building is sited in a hurricane-prone region less than one mile from the coastal mean high water level. The basic wind speed is 110 mph and the glazing is not designed to resist wind-borne debris impact. Thus, as noted earlier, the building must be classified as partially enclosed, irrespective of the openings in the walls and the roof, and the internal pressure coefficients, from Table 6-7 are:

$$(GC_{pi}) = +0.55 \text{ and}$$
$$(GC_{pi}) = -0.55$$

Typical Calculations of Design Pressures for MWFRS (wind parallel to slope with 15 ft windward wall)

Pressure on Leeward Wall
$$p = q_h GC_p - q_h(\pm GC_{pi})$$
$$= 16.32(0.85)(-0.5) - (16.32)(+0.55)$$
$$= 15.9 \text{ psf with positive internal pressure}$$
$$\text{and}$$
$$= 16.32(0.85)(-0.5) - (16.32)(-0.55)$$
$$= 2.0 \text{ psf with negative internal pressure}$$

Pressure on Overhang Top Surface
$$p = q_h GC_p$$
$$= 16.32(0.85)(-0.74)$$
$$= -10.3 \text{ psf}$$

Pressure on Overhang Bottom Surface
(same as windward wall external pressure)
$$p = q_z GC_p$$
$$= 15.01(0.85)(0.8)$$
$$= 10.2 \text{ psf}$$

Note that q_z was evaluated for $z = 15$ ft for bottom surface of overhang as C_p coefficient is based on induced pressures at top of wall.

Table 3.9.4 Design Pressures for MWFRS: Wind Parallel to Roof Slope (normal to ridge line)

Wind Direction	Surface	Z Ft	q_z, q_h psf	Gust Effect, G	External C_p *	Design Pressure, psf $+(GC_{pi})$	Design Pressure, psf $-(GC_{pi})$
Windward Wall 15 ft	Windward wall	0-15	15.01	0.85	0.80	1.2	19.2
	Leeward wall	0-25	16.32	0.85	-0.50	-15.9	2.0
	Side wall	All	16.32	0.85	-0.70	-18.7	-0.7
	Roof	-	16.32	0.85	-0.74	-19.2	-1.3
	Overhang top	-	16.32	0.85	-0.74	-10.3**	-10.3**
	Overhang bottom	-	15.01	0.85	0.80	10.2**	10.2**
Windward Wall 25 ft	Windward wall	0-15	15.01	0.85	0.80	1.2	19.2
		15-20	16.32	0.85	0.80	2.1	20.1
		20-25	17.38	0.85	0.80	2.8	20.8
	Leeward wall	all	16.32	0.85	-0.50	-15.9	2.0
	Side wall	all	16.32	0.85	-0.70	-18.7	-0.7
	Roof	-	16.32	0.85	-0.50	-15.9	2.0
	Overhang top	-	16.32	0.85	-0.50	-6.9**	-6.9**
	Overhang bottom	-	-		-	0.0**	0.0**

* External pressure calculations include G=0.85

** Overhang pressures are not affected by internal pressures. Standard is silent on bottom surface pressures for leeward overhang. One could argue that leeward wall pressure coefficients be applied, but note that neglecting the bottom overhang pressures would be conservative in this application.

Figures 3.9.2 and 3.9.3 illustrate the external, internal, and combined pressure for wind directed normal to 15 ft wall. Figures 3.9.4 and 3.9.5 illustrate combined pressure for wind directed normal to 25 ft wall and perpendicular to slope (parallel to ridge line), respectively.

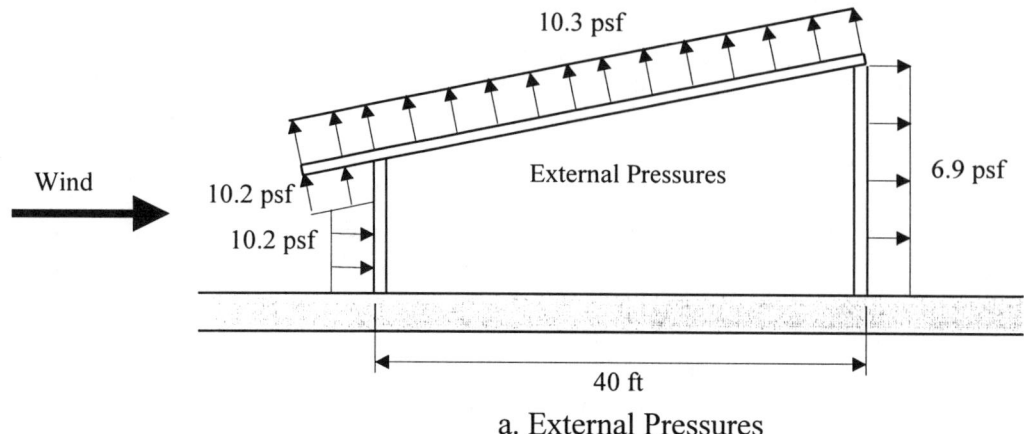

a. External Pressures

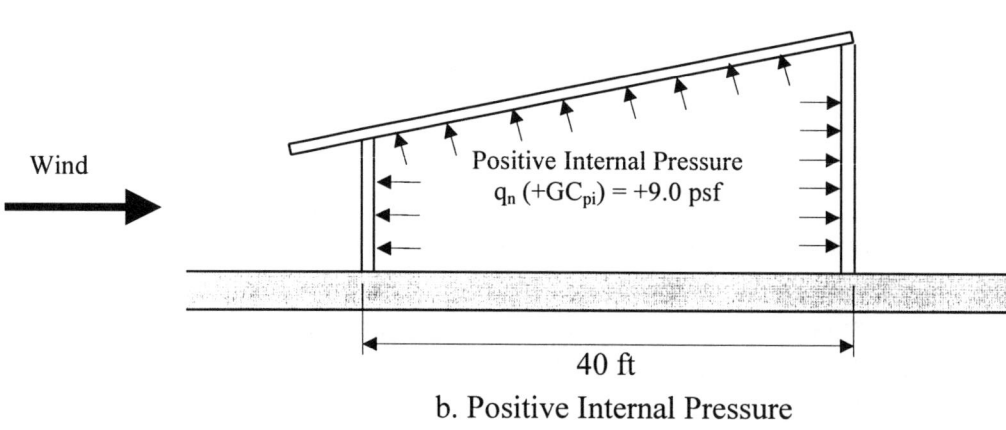

b. Positive Internal Pressure

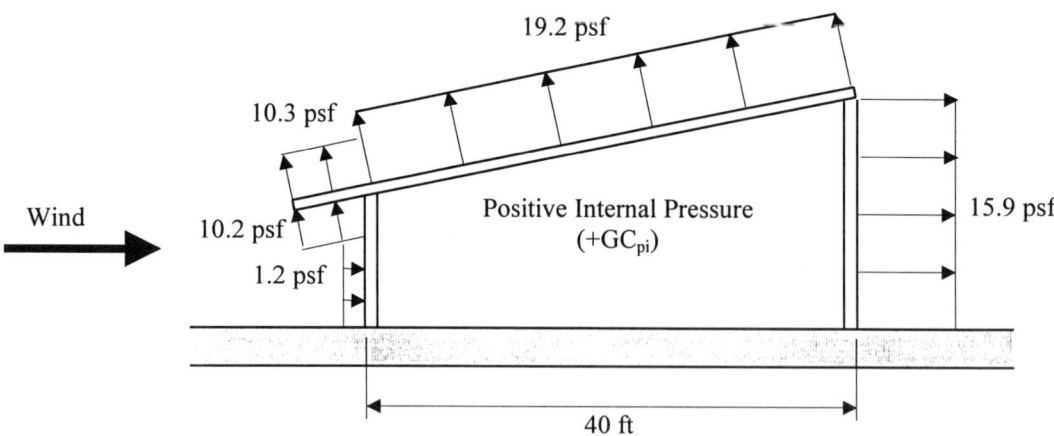

c. Combined External and Positive Internal Pressure

Figure 3.9.2 Design Pressures for MWRFS; Wind Parallel to Roof Slope, Normal to 15 ft Wall and Positive Internal Pressure

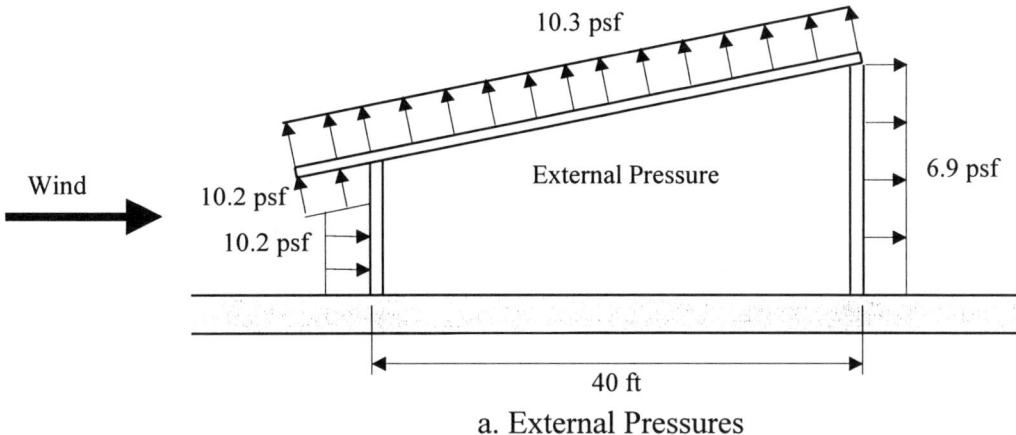

a. External Pressures

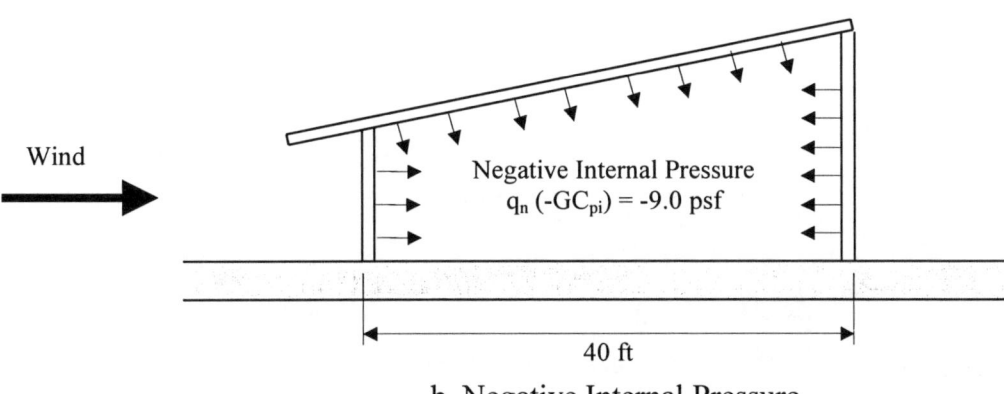

b. Negative Internal Pressure

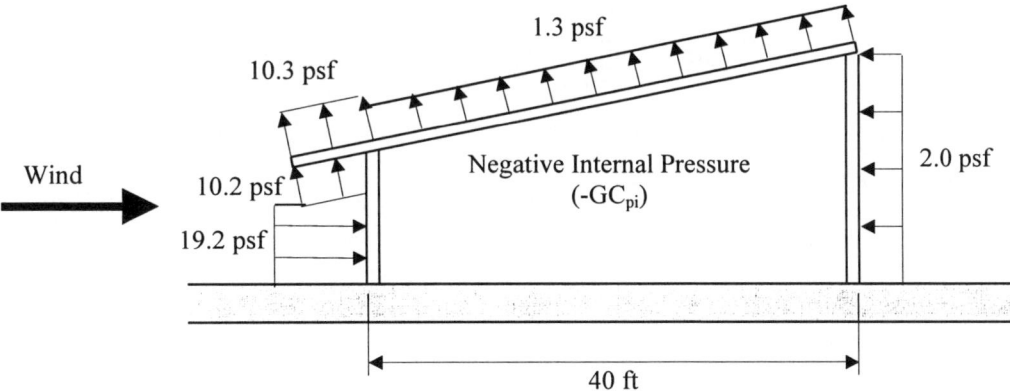

c. Combined External and Negative Internal Pressure

Figure 3.9.3 Design Pressures for MWRFS; Wind Parallel to Roof Slope, Normal (Internal Pressure) to 15 ft Wall and Negative Internal Pressure

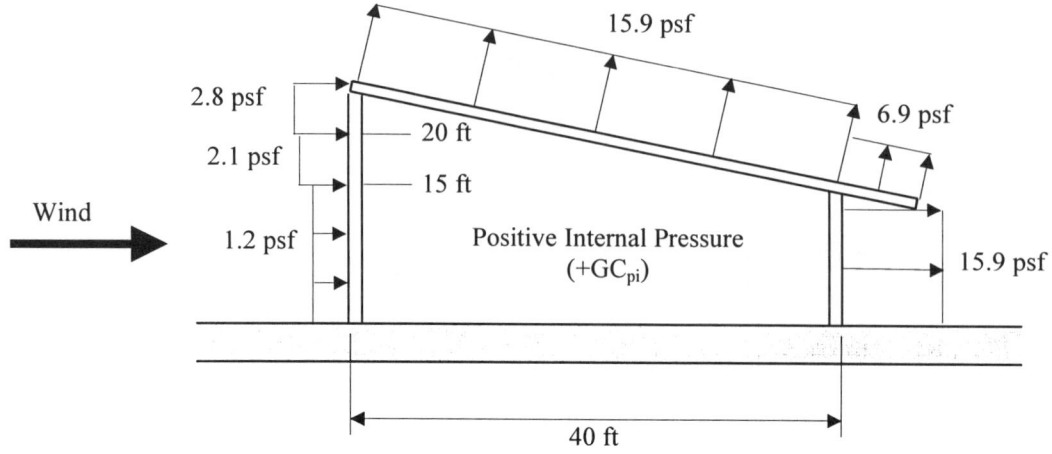

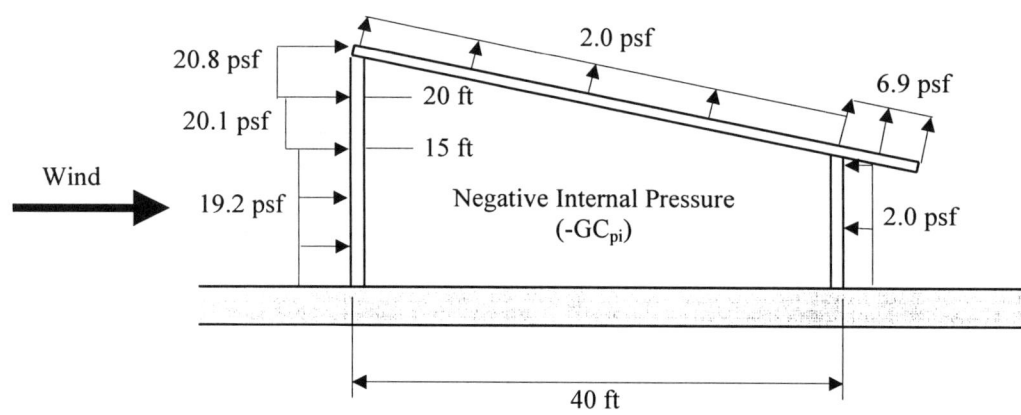

Figure 3.9.4 Combined Design Pressures for MWRFS; Wind Parallel to Roof Slope (Normal to Ridge Line)

Table 3.9.5 Design Pressures for MWFRS: Wind Normal to Roof Slope (parallel to ridge line)

Surface	z or Distance[***] ft	q_z, q_h[*] psf	Gust Effect, G	C_p	Design Pressure, psf	
					$+(GC_{pi})$[**]	$-(GC_{pi})$[**]
Windward wall	0-15	15.01	0.85	0.8	1.2	19.2
	15-20	16.32	0.85	0.8	-2.1	20.1
	20-25	17.38	0.85	0.8	-2.8	20.8
Leeward wall	All	16.32	0.85	-0.3	-13.1	4.8
Side wall	All	16.32	0.85	-0.7	-18.7	-0.7
Roof[****]	0-20	16.32	0.85	-0.9	-21.5	-3.5
	20-40	16.32	0.85	-0.5	-15.9	2.0
	40-80	16.32	0.85	-0.3	-13.1	4.8

[*] External pressure calculations include G = 0.85
[**] Internal pressure is associated with q_h = 16.32 psf
[***] Distance along roof is from leading windward edge
[****] Pressure on overhang is only external pressure (contribution on underside is conservatively neglected)

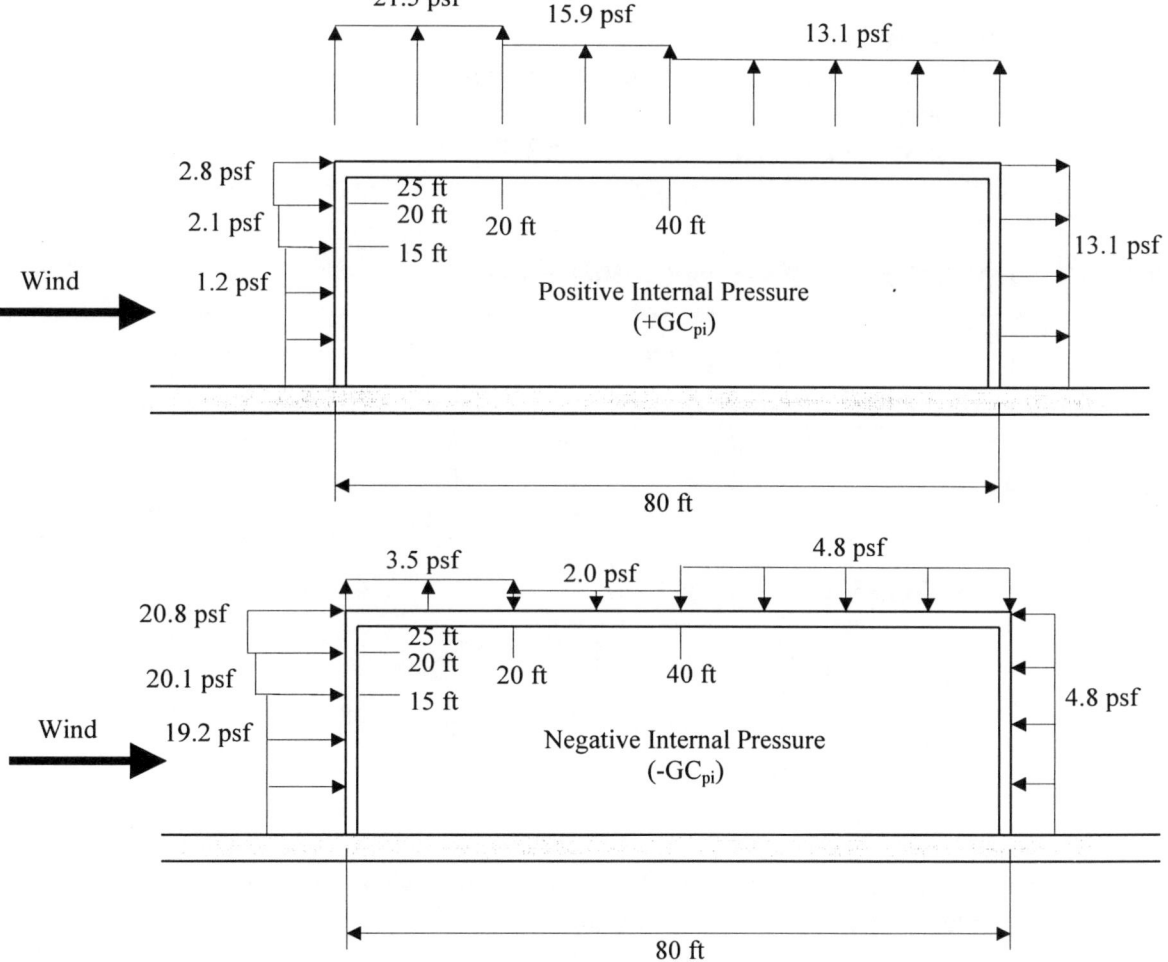

Figure 3.9.5 Combined Design Pressures for MWRFS; Wind Perpendicular to Roof Slope (Parallel to Ridge Line)

Design Pressures for Components and Cladding

Design pressure equation for components and cladding for building with mean roof height h ≤ 60 ft is given in Section 6.5.12.4.1 by Equation 6-18:

$$P = q_h[(GC_p) - (GC_{pi})]$$

where q_h is velocity pressure at mean roof height associated with Exposure B, Case 1 (q_h = 18.43 psf previously determined); (GC_p) are external pressure coefficients from Figure 6-5a, Figure 6-5b, and Figure 6-7a; (GC_{pi}) = +0.55 and -0.55 previously determined from Table 6-7.

Wall Design Pressures
Since the CMU walls are supported at the top and bottom, the effective wind area will depend on the span length.

Effective wind area:
For span of 15 ft, A = 15(15/3) = 75 sq ft
For span of 20 ft, A = 20(20/3) = 133 sq ft
For span of 25 ft, A = 25(25/3) = 208 sq ft

Table 3.9.6 Wall External Pressure Coefficients (GC_p)

| Area A | Pressure Coefficients | | |
sq ft	Zones 4 and 5 (+GC_p)	Zone 4 (-GC_p)	Zone 5 (-GC_p)
75	0.85	-0.95	-1.09
133	0.80	-0.90	-1.00
208	0.77	-0.87	-0.93

Width of Zone 5, Figure 6-5a:

$$a \le \begin{cases} 0.1(40) = 4 \text{ ft (controls)} \\ 0.4(20) = 8 \text{ ft} \end{cases}$$

but not less than

$$a \ge \begin{cases} 0.04(40) = 1.6 \text{ ft} \\ 3 \text{ ft} \end{cases}$$

Design pressures are the critical combinations when the algebraic sum of the external and internal pressures is a maximum.

Typical calculations for Design Pressures for 15 ft Wall, Zone 4

$$p = q_h[(GC_p) - (\pm GC_{pi})]$$
$$= 18.43[(0.85) - (-0.55)]$$
$$= 25.8 \text{ psf}$$
$$\text{and}$$
$$= 18.43[(-0.95) - (0.55)]$$
$$= -27.6 \text{ psf}$$

Table 3.9.7 Wall Design Pressures, psf

Wall Height	Design Pressures, psf		
ft	Zones 4 and 5 Positive	Zone 4 Negative	Zone 5 Negative
15	25.8	-27.6	-30.2
20	24.9	-26.7	-28.6
25	24.3	-26.2	-27.3

Notes: $q_h = 18.43$ psf

The CMU walls should be designed for pressures determined for Zones 4 and 5 using appropriate tributary areas.

The design pressures for doors and glazing can be assessed by using appropriate pressure coefficients associated with their effective wind areas.

Roof Design Pressures
    Effective wind area:
        Roof joist, $A = (41.2)(5) = 206$ sq ft
                or $= (41.2)(41.2/3) = 566$ sq ft (controls)

        Roof panel, $A = (5)(2) = 10$ sq ft (controls)
                Or $= (5)(5/3) = 8.3$ sq ft

Had the effective wind area of the roof joist been greater than 700 sq ft, its external pressure coefficients (GC_p) would still have been determined on the basis of components and cladding. The statement in Section 6.5.12.1.3, in which provisions for MWFRS may be used for a major component is valid only when the tributary area is greater than 700 sq ft. The tributary area for the roof joist is 242 sq ft.

Section 6.5.11.4.2 requires that pressure coefficients for components and cladding of roof overhangs be obtained from Figure 6-5b. Note that the zones for roof overhangs in Figure 6-5b are different from the zones for a monoslope roof in Figure 6-7a.

Table 3.9.8 Roof External Pressure Coefficients (GC$_p$), $\theta = 14°$

Component	Area A	Pressure Coefficient, Figure 6-7a			
	sq ft	Zones 1, 2, and 3 (+GC$_p$)	Zone 1 (-GC$_p$)	Zone 2 (-GC$_p$)	Zone 3 (-GC$_p$)
Joist	566	0.3	-1.1	-1.2	-2.0
Panel	10	0.4	-1.3	-1.6	-2.9
		Pressure Coefficient, Figure 6-5b			
		Zones 1, 2, and 3 (+GC$_p$)	Zone 1 (-GC$_p$)	Zone 2 (-GC$_p$)	Zone 3 (-GC$_p$)
Joist	566	0.3	-0.8	-2.2	-2.5
Panel	10	0.5	-0.9	-2.2	-3.7

Width of zone distance a:

$$a \leq \begin{cases} 0.1(40) = 4 \text{ ft (controls)} \\ 0.4(20) = 8 \text{ ft} \end{cases}$$

but not less than

$$a \geq \begin{cases} 0.04(40) = 1.6 \text{ ft} \\ 3 \text{ ft} \end{cases}$$

The widths and lengths of Zones 2 and 3 for a monoslope roof are shown in Figure 6-7a (they vary from a to 4a), and for overhangs in Figure 6-5b.

Similar to the determination of design pressures for walls, the critical design pressures for roofs are the algebraic sum of the external and internal pressures. The design pressures for overhang areas are based on pressure coefficients obtained from Figure 6-5b.

Typical calculations for Joist Pressures

Zone 2:
$p = q_h[(GC_p) - (\pm GC_{pi})]$
  $= 18.43[(0.3) - (-0.55)]$
  $= 15.7 \text{ psf}$
  and
  $= 18.43[(-1.2) - (0.55)]$
  $= -32.2 \text{ psf}$

Table 3.9.9 Roof Design Pressures, psf

Component	Design Pressures, psf			
	Zones 1, 2, and 3[*] Positive	Zone 1 Negative	Zone 2 Negative	Zone 3 Negative
Joist	15.7	-30.4	-32.2	-47.0
Joist Overhang	10.0[**]	-14.7	-40.6	-46.1
Panel	17.5	-34.1	-39.6	-63.6
Panel in Overhang	10.0[**]	-16.6	-40.6	-68.2

Note: q_h = 18.43 psf

* zones for overhang are in accordance with Figure 6-5b

** Section 6.1.4.2 requires minimum of 10 psf.

Zones for the monoslope roof and for overhang are shown in Figure 4-6. The panels should be designed for the pressures indicated.

Roof joist design pressures need careful interpretation. The high pressures in corner or eave areas do not occur simultaneously at both ends. Two loading cases: wind loadings 1, 2 for joist 1 and wind loadings 3, 4 for joist 2, are shown in Figure 4-6 based on the following zones:

- Joist 1, loading 1: Zones 2 and 3 for roof and Zone 2 for overhang
- Joist 1, loading 2: Zone 2 for roof and Zones 2 and 3 for overhang
- Joist 2, loading 3: Zones 1 and 2 for roof and Zone 1 for overhang
- Joist 2, loading 4: Zone 1 for roof and Zones 1 and 2 for overhang

For simplicity, only one zone is used for overhang pressures in Figure 3-23.

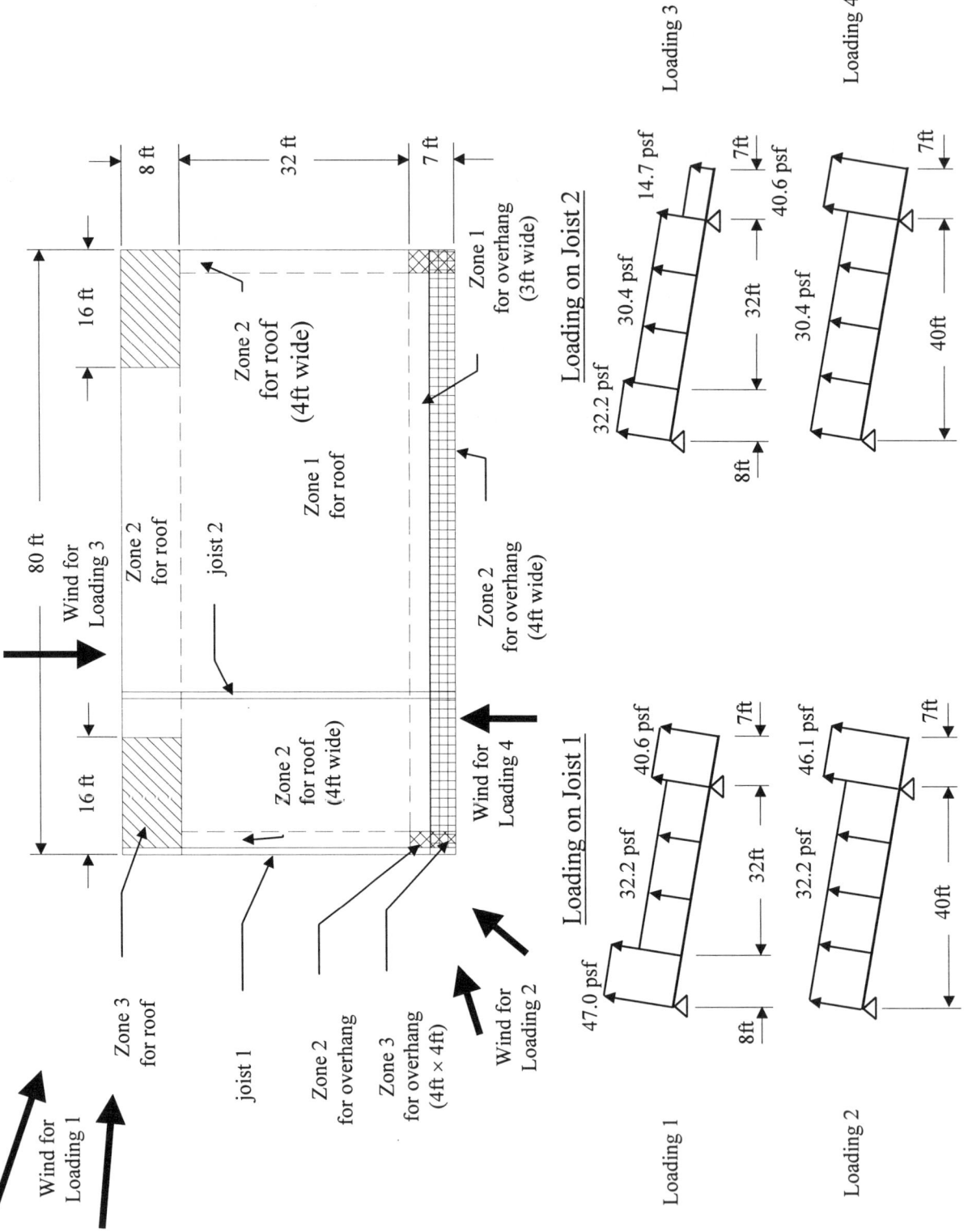

Figure 3.9.6 Design Pressures for Typical Joists and Pressure Zones for Roof Components and Cladding

108

3.10 Example 10 – 50 ft x 20 ft Billboard Sign on Poles (Flexible) 60 ft Above Ground

In this example, design wind-forces for a tall billboard solid sign are determined. The example illustrates two items: (1) determination of G_f for a flexible structure, and (2) use of force coefficient for other structures. The billboard sign data are as follows:

Location: Interstate Highway in Iowa

Terrain: Flat and open terrain

Dimensions: 50 ft x 20 ft sign mounted on two 16 in. diameter steel pipe supports; bottom of the sign is 60 ft above ground.

Structural Characteristics: Tall flexible structure; estimated fundamental frequency is 0.7 Hz and critical damping ratio is 0.01.
(The natural frequency of a structure can be calculated in different ways. It has been predetermined for this example.)

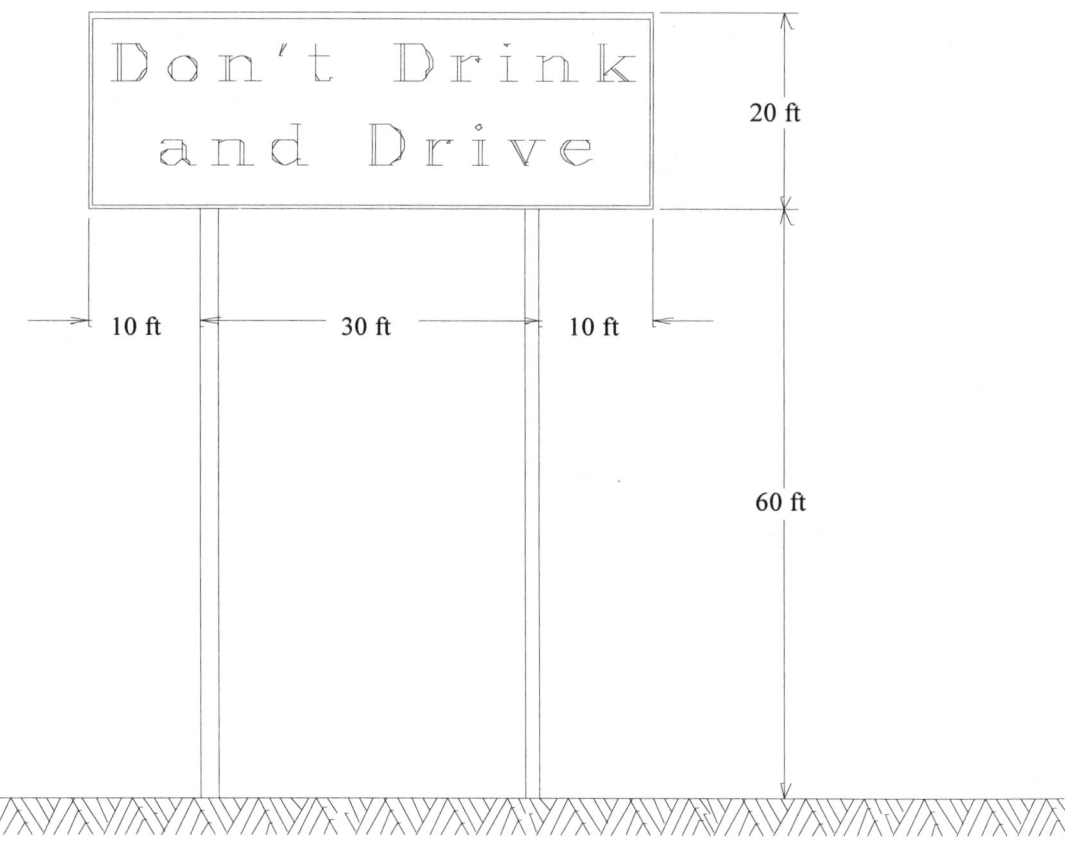

Figure 3.10.1 Dimensions of a Billboard Sign on an Interstate Highway

Exposure and Building Classification

The sign is in an open area; therefore, use Exposure C.
(Section 6.5.6)

Failure of sign represents low hazard to human life since it is
    located away from the highway and is not in populated
    area; the structure can be classified as Category I.   (Table 1-1)

Basic Wind Speed

The wind speed map has only one value of wind speed in the
    middle of the country. Exact location of the sign in Iowa is
    not important. The basic wind speed V = 90 mph.   (Figure 6-1)

Velocity Pressures

The velocity pressures are computed using:

$$q_z = 0.00256\, K_z\, K_{zt}\, K_d\, V^2\, I \text{ psf}$$ (Equation 6-13)

where   V = 90 mph
        I = 0.87 for Category I, see Table 6-1
        K_{zt} = 1.0 because of flat terrain
        K_d = 0.85 for solid sign, see Table 6-6
        K_z = values from Table 6-5 for z of 30, 60 and 80 ft. More
                    divisions of z are not justified because loads on pipe
supports   are small compared to the ones on sign.

Table 3.10.1 Velocity Pressures, psf

Height, ft	K_z	q_z, psf
30	0.98	15.0
60	1.13	17.3
80	1.21	18.6

Design Force for MWFRS

The design force for the MWFRS is given by:

$$F = q_z G_f C_f A_f \qquad \text{(Equation 6-20)}$$

where q_z is as determined previously

G_f is a gust effect factor to be calculated by Equation 6-6 because f < 1 Hz.

$A_f = 50 \times 20 = 1000$ sq ft; for normal and oblique wind, see Note 4 in Table 6-11.

C_f is force coefficient values from Tables 6-10 and 6-11.

Force Coefficient C_f

This sign qualifies as an above ground level sign. (Table 6-11)

$$\frac{M}{N} = 2.5$$

$$C_f = 1.2$$

The supports are round. From Table 6-10:

$$D\sqrt{q_z} = 1.33\sqrt{15.0} = 5.2 > 2.5 \text{ and}$$

$$\frac{h}{D} = \frac{60}{1.33} = 45$$

For moderately smooth surface,

$$C_f = 0.7$$

Gust Effect Factor G_f

The gust effect factor G_f is determined from Equation 6-6:

$$G_f = 0.925 \left[\frac{1 + 1.7 I_{\bar{z}} \sqrt{g_Q^2 Q^2 + g_R^2 R^2}}{1 + 1.7 g_v I_{\bar{z}}} \right]$$

where $I_{\bar{z}}$ is from Equation 6-3

g_Q, g_v are taken as 3.4, see Section 6.5.8.2

g_R is from Equation 6-7

Q is determined from Equation 6-4

R is determined from Equation 6-8

$\bar{z}$ is the equivalent height of the structure, it is used to determine nominal value of $I_{\bar{z}}$; for buildings the recommended value is 0.6h, but for the sign, it should be the middle of the billboard area or 70 ft

c, l, $\bar{\in}$ are given in Table 6-4

$$I_{\bar{z}} = c\left(\frac{33}{\bar{z}}\right)^{1/6} = 0.2\left(\frac{33}{70}\right)^{1/6} = 0.176 \qquad \text{(Equation 6-3)}$$

$$L_{\bar{z}} = l\left(\frac{\bar{z}}{33}\right)^{\bar{\epsilon}} = 500\left(\frac{70}{33}\right)^{1/5} = 581\,\text{ft} \qquad \text{(Equation 6-5)}$$

$$Q^2 = \frac{1}{1 + 0.63\left[\dfrac{B+h}{L_{\bar{z}}}\right]^{0.63}} \qquad \text{(Equation 6-4)}$$

Note: In Equation 6-4, B and h are
the dimensions of the sign.

$$= \frac{1}{1 + 0.63\left[\dfrac{50+20}{581}\right]^{0.63}} = 0.858$$

$$\overline{V}_{\bar{z}} = \overline{b}\left(\frac{\bar{z}}{33}\right)^{\bar{\alpha}} V\left(\frac{88}{60}\right) = 0.65\left(\frac{70}{33}\right)^{1/6.5}(90)\left(\frac{88}{60}\right) = 96.3 \qquad \text{(Equation 6-12)}$$

Note: V is the basic (3-second gust)
wind speed in mph.

$$N_1 = \frac{n_1 L_{\bar{z}}}{\overline{V}_{\bar{z}}} = \frac{(0.7)(581)}{96.3} = 4.22 \qquad \text{(Equation 6-10)}$$

Note: n_1 is the fundamental
frequency of the structure.

$$R_n = \frac{7.47N_1}{(1+10.3N_1)^{5/3}} = 0.0564 \qquad \text{(Equation 6-9)}$$

For R_h, $\eta = \dfrac{4.6n_1h}{\overline{V}_{\bar{z}}} = \dfrac{(4.6)(0.7)(80)}{96.3} = 2.675$

Note: h is taken as 80 ft because resonance response depends on full height.

$$R_h = \frac{1}{\eta} - \frac{1}{2\eta^2}\left(1 - e^{-2\eta}\right) = 0.3043 \qquad \text{(Equation 6-11a)}$$

For R_B (assuming B = 50 ft),

$$\eta = \frac{4.6n_1B}{\overline{V}_{\bar{z}}} = \frac{(4.6)(0.7)(50)}{96.3} = 1.672 \qquad \text{(Equation 6-11a)}$$

$$R_B = \frac{1}{\eta} - \frac{1}{2\eta^2}\left(1 - e^{-2\eta}\right) = 0.4255$$

For R_L (assuming depth L = 2 ft),

$$\eta = \frac{15.4\,n_1L}{\overline{V}_{\bar{z}}} = \frac{(15.4)(0.7)(2)}{96.3} = 0.2239 \qquad \text{(Equation 6-11a)}$$

$$R_L = \frac{1}{\eta} - \frac{1}{2\eta^2}\left(1 - e^{-2\eta}\right) = 0.8661$$

$$g_R = \sqrt{2\ln(3600n_1)} + \frac{0.577}{\sqrt{2\ln(3600n_1)}} \qquad \text{(Equation 6-7)}$$

$$g_R = 4.1$$

$$R^2 = \frac{1}{\beta} R_n R_h R_B (0.53 + 0.47 R_L)$$
(Equation 6-8)

$$= \frac{1}{0.01}(0.0564)(0.3043)(0.4255)[0.53 + (0.47)(0.8661)]$$

$$R^2 = 0.684$$

$$G_f = 0.925 \left[\frac{1 + 1.7 I_{\bar{z}} \sqrt{g_Q^2 Q^2 + g_R^2 R^2}}{1 + 1.7 g_v I_{\bar{z}}} \right]$$
(Equation 6-6)

$$G_f = 0.925 \left[\frac{1 + 1.7(0.176)\sqrt{(3.4)^2(0.858) + (4.1)^2(0.684)}}{1 + 1.7(3.4)(0.176)} \right]$$

$$G_f = 1.093$$

Design Force

Force, $F = q_z G_f C_f A_f$

For one support:
0 to 30 ft	$F = 15.0(1.093)(0.7)(1.33) = 15.3$ plf
30 to 60 ft	$F = 17.3(1.093)(0.7)(1.33) = 17.7$ plf

For two supports:
0 to 30 ft	$F = 30.6$ plf
30 to 60 ft	$F = 35.4$ plf

For a 1 ft horizontal strip of the sign:
$$F = 18.6(1.093)(1.2)(50) = 1220 \text{ plf}$$

The force on the sign follows two cases:
1. Force at geometric center
2. Force at $0.2(50) = 10$ ft on either side of geometric center, see Note 4, Table 6-11

Limitation

In certain circumstances for circular members, across-wind response due to vortex shedding can be critical. The Standard does not provide a procedure to assess across-wind response, but suggests obtaining guidance from recognized literature, see Section 6.5.2.

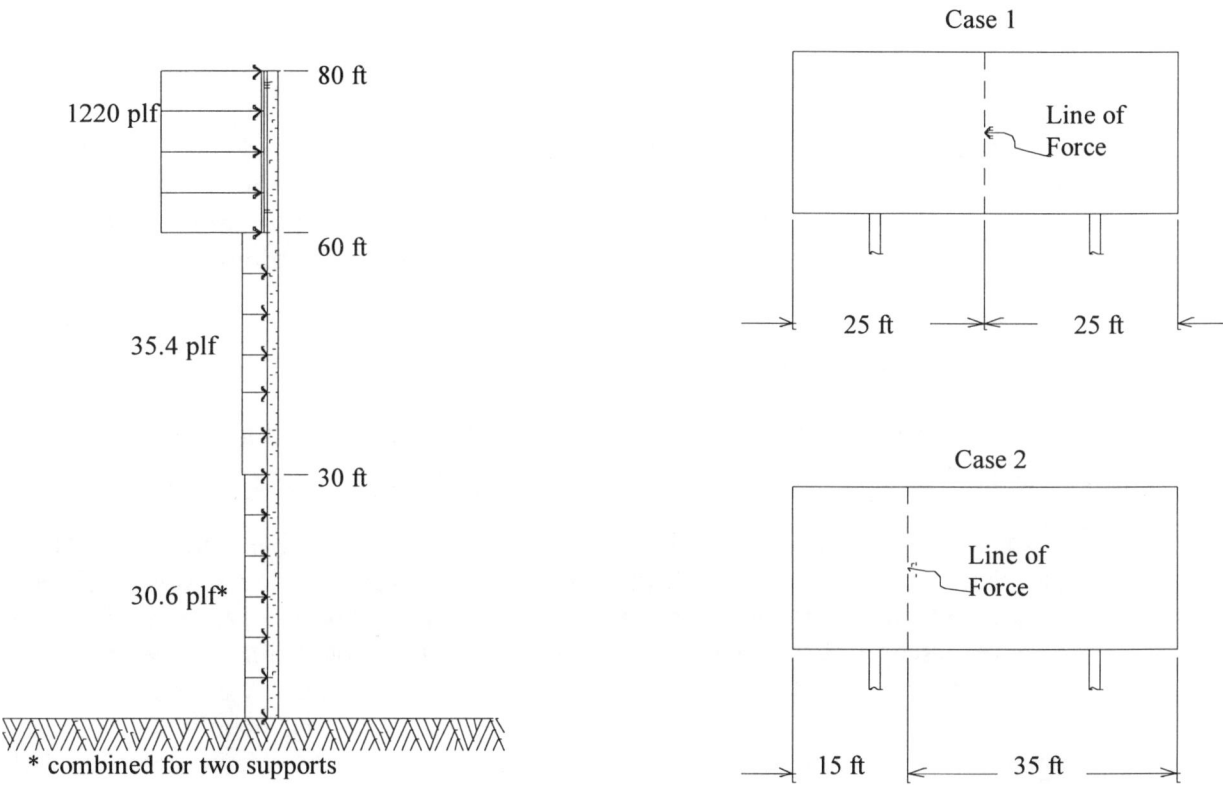

Figure 3.10.2 Design Forces for the Billboard Sign

Force on Components and Cladding

Equation 6-20 is

$$F = q_z G_f C_f A_f$$

The values of q_z are the same as MWFRS except the value of $G = 0.85$. The design forces can be determined using appropriate C_f and A_f for each component or cladding panel.

Chapter 4.
FREQUENTLY ASKED QUESTIONS

4.1   INTRODUCTION

Given the necessary limitations in space in writing a standard and commentary, it is not possible to address all of the design problems that may arise for a variety of the buildings and structures encountered in practice. The purpose of this chapter is to attempt to clarify certain provisions of the standard through frequently asked questions.

4.2   SOME FREQUENTLY ASKED QUESTIONS

1.   When is a gable truss in a house part of the MWFRS? Should it also be designed as a C&C? What about individual members of a truss?

Roof trusses are considered to be components since they receive load directly from the cladding. However, trusses may be designed as part of the MWFRS if they support a tributary area greater than 700 sq ft, see Section 6.5.12.1.3. Even when designed as part of the MWFRS, the top chord of a gable truss may receive load directly from the roof sheathing and thus individual chord and web members may have to be checked using the rules for C&C. When designing shear walls or cross-bracing for light-frame construction such as a house, roof loads can be considered as MWFRS.

2.   For the design of a 4 ft height parapet, what pressure coefficients from the standard can be used for the parapet?

Although the Standard does not provide specific design pressures for parapets, the values of GC_p given in Figure 6-8 allow for rational assessment of these pressures if it is assumed that the parapet acts as a component. For a parapet on the windward building face, the net pressure is assumed to be algebraic sum of the positive pressure for walls, Zones 4 and 5, and the negative pressure for roof, Zone 2. The leeward parapet would receive the positive wall pressure on the back surface (roof side) with the negative wall pressure on the front surface (exterior side of building). There should be no reduction in the positive wall pressure to the leeward parapet due to shielding by the windward parapet, since typically the leeward and windward parapets are too far apart to experience this effect. Since all parapets would be designed for all wind directions, each parapet would in turn be the windward and leeward parapet. The effective wind area will depend upon the dimensions and framing details of the parapet.

For large parapets, MWFRS wind loads can be determined on the basis of the force coefficients C_f given for solid freestanding walls and solid signs in Table 6-11. The coefficients should be based on "at ground level" with the velocity pressure defined by q_z at the height of the parapet. The loads determined by this approach represent the net force on the parapet and include consideration of pressures from both the exterior and interior surfaces.

3.   For the determination of MWFRS loading, what pressure coefficients should be used to reflect contributions for the underside (bottom) of the roof overhangs and balconies?

The MWFRS for windward roof overhangs shall be designed for a positive pressure on the bottom surface corresponding to $C_p = 0.8$ in combination with the pressures indicated in Figures 6-3 and 6-4, Section 6.5.11.4.1, for the top surfaces. No specific guidance is given for balconies, but use of the loading criteria for roof overhangs should be adequate. For components and cladding, roof overhang pressures shall be determined from pressure coefficients given in Figure 6-5b.

4.   What constitutes an open building? If a process plant has a three-story frame with no walls but with a lot of equipment inside the framing, is this an open building?

An open building is a structure having all walls at least 80% open, see Section 6.2. Yes, this three-story frame could be classified as an open building or other structure. In calculating the wind-force, F, appropriate values of C_f and A_f would have to be assigned to the frame and to the equipment inside.

5.   If the mean roof height h is greater than 60 ft with roof geometry represented by Figures 6-5b, 6-6, 6-7a or 6-7b, can the pressure coefficients of these figures be used for roof component and cladding design loads?

Yes, provided the mean roof height is less than 90 ft, the height-to-width ratio does not exceed unity, and q is taken as q_h.

6.   If high winds are accompanied by rain, will the presence of raindrops increase the mean density of the air to the point where the wind loads are affected?

No, although raindrops will increase the mean density of the air, the increase is small and may be neglected. For example, if the average rate of rainfall is 5 ins. per hour, the average density of raindrops will only increase the mean air density by less than 1%.

7.   A masonry wall is 12 ft in height and 80 ft long. It is supported at the top and at the bottom. What effective wind area should be used in determining the design pressure for the wall?

For a given application, the magnitude of the pressure coefficient, GC_p, increases with decreasing effective wind area. Therefore, a very conservative approach would be to consider an effective wind area with a span of 12 ft and a width of 1 ft, and design the wall element as C&C. However, the definition of effective wind area states that this area is the span length multiplied by an effective width that need not be less than one-third the span length. Accordingly, the effective wind area would be $(12)(12/3) = 48$ sq ft.

8.   Roof trusses are 30 ft long and are spaced on 4 ft centers. What effective wind area should be used to determine the design pressures for the trusses?

Roof trusses are classified as components and cladding since they receive wind load directly from the cladding (roof sheathing). In this case, the effective wind area is the span length multiplied by an effective width that need not be less than one-third the span length or $(30)(30/3) = 300$ sq ft. This is the area on which the selection of GC_p should be based. Note however, the resulting wind pressure acts on the tributary area of each truss which is $(30)(4) = 120$ sq ft.

9.   Roof trusses have a clear span of 70 ft and are spaced 8 ft on center. What effective wind area should be used to determine the design pressures for the trusses?

Following the approach of the previous example, the effective wind area is $(70)(70/3) = 1633$ sq ft. The tributary area of the truss is $(70)(8) = 560$ sq ft which is less than the 700 sq ft area required by Section 6.5.12.1.3 to qualify for design of the truss using the rules for MWFRS. The truss is to be designed using the rules for C&C, and the wind pressure corresponding to an effective wind area of 1633 sq ft is to be applied to the tributary area of 560 sq ft.

10.   Metal decking consisting of panels 20 ft long and 2 ft wide is supported on purlins spaced 5 ft apart. Will the effective wind area be 40 sq ft for the determination of pressure coefficients?

Although the length of a decking panel is 20 ft, the basic span is 5 ft. According to the definition of effective wind area, this area is the span length multiplied by an effective width that need not be less than one-third the span length. This gives a minimum effective wind area of $(5)(5/3) = 8.3$ sq ft. However, the actual width of a panel is 2 ft, making the effective wind area equal to the tributary area of a single panel or $(5)(2) = 10$ sq ft. Therefore, GC_p would be determined on the basis of 10 sq ft effective wind area, and the corresponding wind load would be applied to a tributary area of 10 sq ft. Note that GC_p is constant for effective wind areas less than 10 sq ft.

11. A tower has a fundamental frequency of 2 Hz, but has a height-to-width ratio of 6. Should the tower be treated as a flexible structure to determine the gust effect factor?

   No. The guideline of height-to-width ratio of 4 or more is intended to save the user of the Standard the trouble of calculating the fundamental frequency in each and every case. The energy in the turbulence spectrum is very small for frequencies above 1 Hz. Hence, a tower with fundamental frequency of 2 Hz will not be dynamically excited.

12. If the design wind loads are to be determined for a building that is located in a special wind region (shaded areas) in Figures 6-1, 6-1a, 6-1b, and 6-1c, what basic wind speed should be used?

   The purpose of the special wind regions in these figures is to alert the user to the fact that there are regions in which wind speed anomalies are known to exist. Wind speeds in these regions may be substantially higher than the speeds indicated on the map, and the use of regional climatic data and consultations with a wind engineer or consulting meteorologist are advised.

13. If a monoslope roof over an open building is virtually flat, what force coefficients from Table 6-9 should be used?

   A requirement for the use of Table 6-9 is that the wind shall be assumed to deviate ± 10 degrees from the horizontal. Accordingly, the values of C_f corresponding to a roof angle of 10 degrees should be used. The wind-forces may be directed either inward or outward, and both cases should be checked.

14. A trussed tower of 10 ft x 10 ft square cross-section consists of structural angles forming basic tower panels 10 ft high. The solid area of the face of one tower panel projected on a plane of that face is 22 sq ft. What force coefficient, C_f, should be used to calculate the wind-force? What would the force coefficient be for the same tower fabricated of rounded members having the same projected solid area? What area should be used to obtain the wind-force per ft of tower height acting: (1) normal to a tower face, and (2) along a tower diagonal?

   The gross area of one panel face is $(10)(10) = 100$ sq ft, and the solidity ratio is $\varepsilon = 22/100 = 0.22$. For a tower of square cross-section, the force coefficient from Table 6-13 is as follows:

$$C_f = (4)(0.22)^2 - (5.9)(0.22) + 4.0 = 2.90$$

For rounded members, the force coefficient may be reduced by the factor

$$(0.51)(0.22)^2 + 0.57 = 0.59$$

Thus, the force coefficient for the same tower constructed of rounded members with the same projected area would be

$$C_f = (0.59)(2.90) = 1.71$$

The area, A_f, used to calculate the wind-force per ft of tower height is $22/10 = 2.2$ sq ft for all wind directions.

15.   Equation 6-13 for velocity pressure uses the subscript z while Figures 6-3, 6-4, 6-5a, 6-5b, 6.6, 6-7a, 6-7b, and 6-8 use subscripts z and h. When is z used and when is h used?

Equation 6-13 is the general formula for the velocity pressure q_z at any height z above ground. There are many situations in the Standard where a specific value of z is called for, namely the height (or mean roof height) of a building or other structure. Whenever the subscript h is called for, it is understood that z becomes h in the appropriate equations.

16.   The Standard does not provide for across-wind excitation caused by vortex shedding. How can one determine when vortex shedding might become a problem?

Vortex shedding is almost always present with bluff-shaped cylindrical bodies. For structures having non-cylindrical cross-sections, it can become a problem when the frequency of shedding is close to or equal to the frequency of the first or second transverse modes of the structure. The intensity of excitation increases with aspect ratio (height-to-width or length-to-breadth) and decreases with increasing structural damping. Structures with low damping and with aspect ratio of 8 or more may be prone to damaging vortex excitation. If across-wind or torsional excitation appears to be a possibility, expert advice should be obtained.

17.   Under what conditions is it necessary to consider speed-up due to topographic effects when calculating wind loads?

The Standard requires the calculation of the topographic factor K_{zt} for buildings and other structures sited on the upper half of isolated hills or escarpments located in Exposures B, C, or D where the upwind terrain is free of such topographic features for a distance of at least 100H or two miles, whichever is smaller, as measured from the crest of the topographic feature. K_{zt} need not be calculated when the height H is less than 15 ft in Exposures D and C, or less than 60 ft in Exposure B. In addition, K_{zt} need not be calculated when H/L_h is less than 0.2. H and L_h are defined in Figure 6-2. The value of K_{zt} is never less than 1.0.

18.   Figure 6-9 shows wind pressure combinations to be used for full and partial loading in the design of main wind-force resisting systems in buildings with mean roof height greater than 60 ft. Does the Standard have similar requirements for buildings with mean roof height less than or equal to 60 ft?

       No. The Standard does not require the designer to consider the partial loadings of Figure 6-9 when designing buildings with mean roof height less than or equal to 60 ft. However, the actions of diagonally-acting resultants and torsional moments are accounted for automatically when using the external pressure coefficients, GC_{pf}, given for load Cases A and B in Figure 6-4.

19.   In the design of main wind-force resisting systems, the provisions of Figure 6-3 apply to enclosed or partially enclosed buildings of all heights. The provisions of Figure 6-4 apply to enclosed or partially enclosed buildings with mean roof height less than or equal to 60 ft. Does this mean that either figure may be used for the design of low-rise MWFRS?

       Figure 6-3 may be used for buildings of all heights, whereas Figure 6-4 applies only to buildings having a gable roof and a mean roof height less than or equal to 60 ft. Thus, for a low-rise building having a gable roof, the choice of approach is left to the designer. It should be noted that the velocity pressure to be used for Figure 6-4 is based on mean roof height and the pressures are assumed to be uniformly distributed over the areas of interest.

20.   Section 6.1.4.1 provides for a minimum wind pressure of 10 lb/ft^2 multiplied by the area of the building or structure projected onto a vertical plane normal to the assumed wind direction for MWFRS. Does this same provision apply to Figure 6-4.

       No. The values for (GC_{pf}) given in Figure 6-4 represent "pseudo" loading conditions, Cases A and B. When applied to the building, these coefficients envelope the induced internal and external structural actions (bending moment, shear, and thrust) independent of wind direction. As such, the wind pressure on a vertical projected plane may be less than 10 lb/ft^2.

21.   In calculating the wind-forces acting on a trussed tower of square cross-section, see Table 6-13, should the force coefficient, C_f, be applied to both the front and the back (windward and leeward) faces of the tower?

No. The calculated wind-forces are the total forces acting on the tower. The force coefficients given in Table 6-13 include the contributions of both front and back faces of the tower as well as the shielding effect of the front face on the back face.

22.   Is it possible to obtain larger scale maps of basic wind speeds, see Figures 6-1, 6-1a, 6-1b, and 6-1c, so that the locations of the wind speed contours can be determined with greater accuracy?

No. The wind speed contours in the hurricane-prone region of the United States are based on hurricane wind speeds from Monte Carlo simulations and on estimates of the rate at which hurricane wind speeds attenuate to 90 mph following landfall. Because the wind speed contours of these figures represent a consensus of the ASCE 7 Wind Load Task Committee, a larger scale map would do nothing to improve their accuracy.

23.   If the pressure or force coefficients for various roof shapes, e.g., a canopy, are not given in ASCE 7-98, how can the appropriate wind-forces be determined for these shapes?

With the exception of pressure or force coefficients for certain shapes, parameters such as V, I, K_z, K_{zt}, and G are given in ASCE 7-98. It is possible to use pressure or force coefficients from the published literature, see Section 1.4, provided these coefficients are used with care. Mean pressure or force coefficients from other sources can be used to determine wind loads for MWFRS. However, it should be recognized that these coefficients may have been obtained in wind tunnels having smooth, uniform flows as opposed to more proper turbulent boundary-layer flows. Pressure coefficients for components and cladding obtained from the literature should be adjusted to the 3-second gust speed which is the basic wind speed adopted by ASCE 7-98.

24.   Can the pressure/force coefficients given in ASCE 7-98 be used with the provisions of ASCE 7-88, 7-93 or 7-95?

Yes, in a limited way. ASCE 7-88 (and 7-93) used the fastest-mile wind speed as the basic wind speed. With the adoption of the 3-second gust speed in ASCE 7-95, the values of certain parameters used in the determination of wind-forces have been changed accordingly. The provisions of ASCE 7-88 and 7-98 should not be interchanged.

25.   Is it possible to determine the wind loads for the design of interior walls?

The Standard is silent on the wind loads to be used in the design of interior walls or partitions. A conservative approach would be to apply the internal pressure coefficients $GC_{pi} = \pm 0.18$ for enclosed buildings and $GC_{pi} = \pm 0.55$ for partially enclosed buildings. Post disaster surveys have revealed the failure of interior walls when the building envelope has been breached.

26. Section 6.2 of the Standard provides definitions of glazing, impact resistant; impact resistant covering; and wind-borne debris regions. To be impact resistant, the Standard specifies the glazing of the building envelope must be shown by an approved test method to withstand the impact of wind-borne missiles likely to be generated during design winds. Where does one find information on appropriate test methods?

A number of code and national standard agencies (e.g., SBCCI, Miami/Dade Building Code Department, ASTM, TDI) have developed appropriate protocols to satisfy the wind-borne debris issue, see Table 1.4.1.

LIST OF REFERENCES

Akins, R.E. and Cermak, J.E. (1975). *Wind pressures on buildings*. Technical Report CER7677REA-JEC15, Fluid Dynamics and Diffusion Laboratory, Colorado State University, Fort Collins, Colorado.

ANSI A58.1-1982 (1982). *Minimum Design Loads for Buildings and Other Structures*. American National Standards Institute, New York, New York.

ANSI/EIA/TIA-222-E (1991). "Structural standards for steel antenna towers and antenna supporting structures," Electronic Industries Assn., Washington, DC.

ASCE (1961). "Wind forces on structures," *Transactions,* ASCE, 126 (2), 1124-1198.

ASCE (1987). "Wind tunnel model studies of buildings and structures," *Manuals and Reports on Engineering Practice,* No. 67, American Society of Civil Engineers, New York, New York.

ASCE (1997). *Wind Loads and Anchor Bolt Design for Petrochemical Facilities*. Report from the Committee on Wind Induced Forces and Anchor Bolt Design. American Society of Civil Engineers, New York, New York.

ASCE/ANSI 7-88 (1990). *Minimum Design Loads for Buildings and Other Structures*. American Society of Civil Engineers, New York, New York.

ASCE/ANSI 7-95 (1996). *Minimum Design Loads for Buildings and Other Structures*. American Society of Civil Engineers, New York, New York.

AS-1170.2 (1989). "Australian Standard SAA Loading Code Part 2," Wind Loads, Standards Australia.

AS/NZS 1170.2, "Australian/New Zealand Structural Design – General Requirements and Design Actions Part 2: Wind Actions," Standards Australia.

ASTM E1886-97 "Standard Test Method for Performance of Exterior Windows, Curtain Walls, Doors and Storm Shutters Impacted by Missile(s) and Exposed to Cyclic Pressure Differentials."

ASTM E1996-01 "Standard Specification for Performance of Exterior Windows, Curtain Walls, Doors and Storm Shutters Impacted by Windborne Debris in Hurricanes."

Batts, M.E., Cordes, M.R., Russell, L.R., Shaver, J.R. and Simiu, E. (1980). "Hurricane wind speeds in the United States," NBS Building Science Series 124, National Bureau of Standards, Washington, DC.

Behr, R.A., and Minor, J.E. (1994). "A survey of glazing system behavior in multi-story buildings during Hurricane Andrew," *The Structural Design of Tall Buildings*, 3, 143-161.

Best, R.J. and Holmes, J.D. (1978). "Model study of wind pressures on an isolated single-story house," Wind Engineering Report 3/78, James Cook University of North Queensland, Australia.

British Standard/BS 6399 (1995). "Loading for Buildings, Part 2: Code of Practice for Wind Loads," British Standards Institute, London W4 4AL

Cermak, J.E. (1977). "Wind-tunnel testing of structures," *Journal of Engineering Mechanics Division.*, ASCE, 103(6), 1125-1140.

Cook, N.J. (1985). *The Designer's Guide to Wind Loading of Building Structures,* Parts I and II, Butterworth Publishers, London, England.

Davenport, A.G., Surry, D. and Stathopoulos, T. (1977). "Wind loads on low-rise buildings," Final Report on Phases I and II, BLWT-SS8, University of Western Ontario, Canada.

Davenport, A.G., Surry, D. and Stathopoulos, T. (1978). "Wind loads on low-rise buildings," Final Report on Phase III, BLWT-SS4, University of Western Ontario, Canada.

Durst, C.S. (1960). "Wind speeds over short periods of time," *Meteorological Magazine,* 89, 181-187.

Eaton, K.J. and Mayne, J.R. (1975). "The measurement of wind pressures on two-story houses at Aylesbury," *Journal of Industrial Aerodynamics,* 1(1), 67-109.

EUROCODE 1 (1994). *Basis of Design and Actions on Structures/Wind Action, Part 2.3: Wind Actions,* CEN/TC 250/SC1, Steering Panel Draft, May 1994.

FEMA (1980). *Interim Guidelines for Building Occupant Protection from Tornadoes and Extreme Winds*, TR83-A, Federal Emergency Management Agency, Washington, DC.

FEMA (1999). *Taking Shelter from the Storm: Building a Saferoom Inside Your Home.* Publication 320, Federal Emergency Management Agency, Washington, DC.

FEMA (2000). *Design and Construction Guidance for Community Shelters.* Publication 361, Federal Emergency Management Agency, Washington, DC.

Georgiou, P.N., Davenport, A.G. and Vickery, B.J. (1983). "Design wind speeds in regions dominated by tropical cyclones," *Journal of Wind Engineering and Industrial Aerodynamics,* 13, 139-152.

Ho, E. (1992). "Variability of low building wind loads," Doctoral Dissertation, University of Western Ontario, London, Ontario, Canada.

Holmes, J.D. (2001). *Wind Loading of Structures.* New York: Spon Press.

Holmes, J.D., Melbourne, W.H., and Walker, G.R. (1990). "A Commentary on the Australian Standard for Wind Loads," Australian Wind Engineering Society (printed by Courtney Colour Graphics Pty. Ltd., Lilydale, Victoria, Australia).

Hoerner, S.F. (1965). *Fluid Dynamics Drag,* published by the author, Midland Park, NJ.

ISO. (1997). "Wind Actions on Structures," ISO 4354, International Organization for Standardization, Geneva, Switzerland.

Isyumov, N. (1982). "The aeroelastic modeling of tall buildings," *Proceedings,* International Workshop on Wind Tunnel Modeling Criteria and Techniques in Civil Engineering Applications, NBS, Gaithersburg, MD, Cambridge University Press, 373-407.

Isyumov, N. and Case, P. (1995). "Evaluation of structural wind loads for low-rise buildings contained in ASCE Standard 7-1995," BLWT-SS17-1995, University of Western Ontario, London, Ontario, Canada.

Jackson, P.S. and Hunt, J.C.R. (1975). "Turbulent wind flow over a low hill," *Quarterly Journal of the Royal Meteorological Society,* 101, 929-955.

Kareem, A. (1992). "Dynamic response of high-rise buildings to stochastic wind loads," *Journal of Wind Engineering and Industrial Aerodynamics,* 41-44.

Kareem, A. (1985). "Lateral-torsional motion of tall buildings to wind loads," *Journal of Structural Engineering,* ASCE 111(11).

Kareem, A. and Smith, C. (1993). "Performance of offshore platforms in Hurricane Andrew," *Proceedings,* Hurricanes of 1992, ASCE, Dec. 1-3, Miami, FL, Dec.

Kavanagh, K.T., Surry, D., Stathopoulos, T. and Davenport, A.G. (1983). "Wind loads on low-rise buildings: Phase IV," BLWT-SS14, University of Western Ontario, London, Ontario, Canada.

Krayer, W.R. and Marshall, R.D. (1992). "Gust factors applied to hurricane winds," *Bulletin of the American Meteorological Society,* Vol. 73, 613-617.

Lawson, T.V. (1980). *Wind Effects on Buildings,* Volumes 1 and 2, Applied Science Publishers Ltd., Ripple Road, Barking, Essex, England

Lemelin, D.R., Surry, D. and Davenport, A.G. (1988). "Simple approximations for wind speed-up over hills," *Journal Wind Engineering and Industrial Aerodynamics,* 28, 117-127.

Liu, Henry (1991). *Wind Engineering: A Handbook for Structural Engineers*, Prentice-Hall, New York, NY.

Marshall, R.D. and Yokel, F.Y. (1995). *Recommended performance-based criteria for the design of manufactured home foundation systems to resist wind and seismic loads,* NISTIR 5664, National Institute of Standards and Technology, Gaithersburg, MD.

McDonald, J.R. (1983). "A Methodology for Tornado Hazard Probability Assessment," NUREG/CR3058, U.S. Nuclear Regulatory Commission, Washington, DC.

Miami/Dade County Building Code Compliance Office Protocol PA 201-94, "Impact Test Procedures."

Miami/Dade County Building Code Compliance Office Protocol PA 203-94, "Criteria for Testing Products Subject to Cyclic Wind Pressure Loading."

Minor, J.E. (1982). "Tornado technology and professional practice," *Journal of the Structural Division,* ASCE, 108(11), 2411-2422.

Minor, J.E. and Behr, R.A. (1993). "Improving the performance of architectural glazing systems in hurricanes," *Proceedings,* Hurricanes of 1992, ASCE, Dec. 1-3, Miami, FL, pp C1-11.

Minor, J.E., McDonald, J.R. and Mehta, K.C. (1977). *The tornado: An engineering oriented perspective*, TM ERL NSSL-82, National Oceanic and Atmospheric Administration, Environmental Research Laboratories, Boulder, CO.

Murray, R.C. and McDonald, J.R. (1993). Design for containment of hazardous materials, *Geophysical Monograph 79,* The Tornado: Its Structure, Dynamics, Prediction and Hazards, C. Church, D. Burgess, C. Doswell and R. Davies-Jones, Eds, American Geophysical Union, 379-387.

Newberry, C. W., and Eaton, K.J. (1974). *Wind Loading Handbook*, Building Research Establishment Report K4F, Her Majesty's Stationery Office, London, England.

NRCC (1995a). *National Building Code of Canada, 1995*, Associate Committee on the National Building Code of Canada, National Research Council of Canada.

NRCC (1995b). *Supplement to the National Building Code of Canada, 1995*, Associate Committee on the National Building Code of Canada, National Research Council of Canada.

Peterka, J.A. (1992). "Improved extreme wind prediction for the United States," *Journal of Wind Engineering and Industrial Aerodynamics,* 41, 533-541.

Peterka, J.A. and Cermak, J.E. (1974). "Wind pressures on buildings-Probability densities," *Journal of the Structural Division,* ASCE, 101(6), 1255-1267.

Peterka, J.A. and Shahid, S. (1993). "Extreme gust wind speeds in the U.S.," *Proceedings,* 7th U.S. National Conference on Wind Engineering, UCLA, Los Angeles, CA, 2, 503-512.

Reinhold, T.A. (Ed.) (1982). "Wind tunnel modeling for civil engineering applications," *Proceedings,* International Workshop on Wind Tunnel Modeling Criteria and Techniques in Civil Engineering Applications, NBS, Gaithersburg, MD, Cambridge University Press.

SAA (1989). *Australian Standard SAA Loading Code, Part 2: Wind Loads,* published by Standards Australia, Standards House, 80 Arthur St., North Sydney, NSW, Australia.

Saathoff, P. and Stathopoulos, T. (1992). "Wind loads on buildings with sawtooth roofs," *Journal of Structural Engineering,* ASCE, 118(2), 429-446.

SBCCI. (1999). *SBCCI Test Standard for Determining Impact Resistance from Windborne Debris.* SSTD 12-99, Southern Building Code Congress International, Birmingham, AL, 1999.

Simiu, E. (1981). "Modern developments in wind engineering: Part 1-4," *Engineering Structures,* 3.

Simiu, E. and Scanlan, R.H. (1996). *Wind effects on structures,* Third Edition, John Wiley & Sons, New York, NY.

Solari, G. (1993). "Gust buffeting I: Peak wind velocity and equivalent pressure," *Journal of Structural Engineering,* ASCE, 119(2).

Solari, G. (1993). "Gust buffeting II: Dynamic along-wind response," *Journal of Structural Engineering,* ASCE, 119(2).

Stathopoulos, T. (1981). "Wind loads on eaves of low buildings," *Journal of the Structural Division,* ASCE, 107(10), 1921-1934.

Stathopoulos, T. and Dumitrescu-Brulotte, M. (1989). "Design recommendations for wind loading on buildings of intermediate height," *Canadian Journal of Civil Engineering,* 16(6), 910-916.

Stathopoulos, T. and Luchian, H.D. (1990). "Wind pressures on building configurations with stepped roofs," *Canadian Journal of Civil Engineering,* 17(4), 569-577.

Stathopoulos, T. and Mohammadian, A.R. (1986). "Wind loads on low buildings with mono-sloped roofs," *Journal of Wind Engineering and Industrial Aerodynamics,* 23, 81-97.

Stathopoulos, T. and Saathoff, P. (1991). "Wind pressures on roofs of various geometries," *Journal of Wind Engineering and Industrial Aerodynamics,* 38, 273-284.

Surry, D. and Stathopoulos, T. (1988). "The wind loading of buildings with monoslope roofs," Final Report, BLWT-SS38, University of Western Ontario, London, Ontario, Canada.

TDI. (1998). TDI Standard 1-98, "Test for Impact and Cyclic Wind Pressure Resistance of Impact Protective Systems and Exterior Opening Systems," Appendix E, *Building Code for Windstorm Resistant Construction.* Austin, Texas: Texas Department of Insurance.

Vickery, P.J. and Twisdale, L.S. (1993). "Prediction of hurricane wind speeds in the U.S.," *Proceedings,* 7th U.S. National Conference on Wind Engineering, UCLA, Los Angeles, CA, 2, 823-832.

Walmsley, J.L., Taylor, P.A. and Keith, T. (1986). "A simple model of neutrally stratified boundary-layer flow over complex terrain with surface roughness modulations," *Boundary-Layer Meteorology,* 36, 157-186.

Wen, Y.K. and Chu, S.L. (1973). "Tornado risks and design wind speed," *Journal of the Structural Division,* ASCE, 99(12), 2409-2421.

Yeatts, B.B. and Mehta, K.C. (1993). "Field study of internal pressures," *Proceedings,* 7th U.S. National Conference on Wind Engineering, UCLA, Los Angeles, CA, 2, 889-897.

Yeatts, B.B., Womble, J.A., Mehta, K.C. and Cermak, J.E. (1994). "Internal pressures for low-rise buildings," *Proceedings,* Second U.K. Conference on Wind Engineering, Warwick, England.

SUBJECT INDEX

Errata for *Guide to the Use of the Wind Load Provisions of ASCE 7-98*

by Kishor C. Mehta and Dale C. Perry

Page 33   Corner Zone 5, p = 31.3 [(0.85) (0.9) – (±0.18)] = +29.7 (3.13 should be 31.3)
Middle Zone 4, p = 31.3 [(0.85) (0.9) – (±0.18)] = +29.7 (3.13 should be 31.3)

Page 34   Eave Zone 2 and Corner Zone 3 (on top half of page)
   p = 31.3 [-0.9 ± 0.18)] = -33.8 psf replace with
   p = 31.3 [-1.1 ± 0.18)] = -40.1 psf

Page 46   In the title for Figure 3.3.2, delete the word "Net."
Add a note to the figure: "Note: Internal pressures of +6.3 psf and –6.3 psf are
   to be added to the pressures shown, resulting in two load cases."

Page 47   In Figure 3.3.3, add the same note as added for Figure 3.3.2 on page 46.

Page 72   Table 3.7.6 Leeward Wall C_p = -0.45 (minus sign is missing)

Page 81 K_{zt} = 1.0 topographic factor, see Section 6.5.7.1 (instead of 6.5.71)

Page 83   Corner 1, Case A: Surface 2E
   p = 18.0 [(-1.07) – (±0.18)] (instead of 21.2; other numbers are correct)

Page 84   Table 3.8.6 Corner 2: Case A
   Building surface 4; $(+GC_{pi})$ = -8.5 (instead of –7.6)

Page 90   Figure 3.8.6 Value at top of the page should be 8.5 psf (instead of 7.6 psf)